Anthropometry for designers

John Croney

Batsford Academic and Educational Limited *London*

15994

065818

Dedicated to Isobel, Mary and Sarah

© John Croney 1980

ISBN 0 7134 1567 3

First published 1971
New edition 1980

IBM set by Tek-Art, London SE20
Printed in Great Britain by
The Anchor Press Ltd
Tiptree, Essex
for the publishers
Batsford Academic and Educational Limited
4 Fitzhardinge Street, London W1H 0AH

Contents

Acknowledgment to first edition

I wish to acknowledge the help I have had from many people and sources in the course of the preparation of this book.

Published anatomical and anthropological information has been the source of some of the facts used, other information has come from experts who have shared their knowledge and experience.

I am grateful to Professor J N Tanner and to his publishers Blackwell Scientific Publications, Oxford, for allowing me to use material from *Growth at Adolescence* as a basis for my own findings for this book. Figure 3 was derived in part from a photograph showing the proportions of boys' faces in the above mentioned work. The idea of the fitted curve showing incremental increases in growth shown in Figures 15a and 15b has been derived from the same source. The measurements have been calculated by me and for the purpose of clarity stated as simple numbers.

I am grateful to Dr W H Sheldon and his publishers Harper and Row, New York, for allowing me to use photographs of male somatotypes from the *Atlas of Men* to construct the examples shown in Figures 27, 28 and 29 and for the use of the information in *The Varieties of Human Physique* on the incidence of somatotypes in a population. This information forms the basis for Table 9.

My thanks are due to my colleagues who have freely made available their ideas, and who have directed my attention towards certain aspects of human measurement I may otherwise have missed. They are due as well to Robert Arnold for the generous and lucid manner with which he shared biological knowledge. Also to A E Halliwell and the teaching staff of the Industrial Design Department at the Central School of Art and Design for allowing me to benefit from their experience gained when designing equipment for a variety of human tasks. I am grateful for advice and specifications given by Nigel Walters on the design of seats and domestic furniture in relation to human measurements and use.

Any errors in the use of the measurements and information I have been allowed to use are entirely my own, as are the conclusions I have reached from them.

In the field of practical research I must thank all the models who patiently assisted my study of human measurement and physique.

I must express my debt to Miss B J Kirkpatrick and the Library staff of the Royal Anthropological Institute; to Mrs K Ely of the Croydon Library service and to Miss A M Coghlan, librarian of the College of Fashion and Clothing Technology who have gone to much trouble to obtain material of use.

The preparation of the manuscript was made much easier by the typing and assistance of Miss E Ridley in the initial stage; and later by Miss M Leach.

My greatest debt is to those at Batsfords who were concerned with the publication of the book and the personal interest they have shown in it: to Guy Warren for the design of the layout; to Miss Mary Scriven who made liaison arrangements so easy and pleasant; and lastly to Samuel Carr for his invaluable guidance, unfailing enthusiasm and kindness.

London   J C 1971

Acknowledgment to second edition

For permission to reproduce tables and figures grateful acknowledgment is made to the publishers and editors of *Man* (NS) 12, Croney, J (1977); the Journal of the Royal Anthropological Institute, and W F F Kemsley, *Women's Measurements and Sizes*: a study sponsored by the Joint Clothing Council Limited published by Her Majesty's Stationery Office.

I must particularly thank Thelma M Nye for her very thorough and painstaking work in editing the numerous new additions and small corrections that were essential in the preparation of this second edition.

London   J C 1980

Introduction

Nothing seems quite so permanent as man in the contemporary situation, despite all we say to the contrary. He seems to be the one constant factor amongst the ever increasing amount of inventions with which he hedges himself in. It is impossible to correlate artistic and industrial efforts to help manufacture things of use, without gathering information about man first of all. If the information is to be worthwhile and apposite one of the things man must do is to measure himself. It is this necessary concern of man with himself, and the tools he wants to use, that is the justification for this book.

To know about man is to know a great deal about the possible environments in which he can function satisfactorily; and in fact the most successful environmental system would be a replica of his own system. That is a system that could function in the same atmosphere, at the same temperature, etc. To be a successful designer of products for a part, or the whole, of a human environment, a designer must be conversant with different types of human physique and be informed about the limitations of human performance. Also now, more than ever before, man demands products that are tailor-made to suit specific requirements.

This book has been designed to provide an illustrated account of man's dimensions and other physical data, and to define their limitations and comment on any peculiarities. It is intended as a reference for use by designers in specialised fields of industrial or commercial design (for example in the furniture and dress trades), and by engineers and architects.

If we consider that the work of many designers is concerned with solving problems of environment it is hoped that the book will help towards the invention of fresh solutions. The contents deal with the physical attributes and shapes of man and are intended only as a first guide to the solutions of specific problems.

One of the more important aspects of this type of book is the consideration of how best to describe the facts and ideas it contains. A great deal of anthropometric information can be conveyed visually; and a scheme consisting of a series of figures and tables was decided upon.

The text has been kept as clear and simple as possible, and the aim has been to achieve a pertinent association of works and diagrams. In this way a course has been steered between a text overloaded with natural-science language, and one that is either too extravagantly personal or too scanty and journalistic.

When the aim is to be useful and comprehensive one cannot be original as well. Many of the facts used are not my own; but they are also not generally available. A great many have been collected from the technical literature of Europe and America, and I have also taken facts from the books listed in the Bibliography. I have incorporated much current thinking in anthropometry, human biology and anatomy and have made the information as up-to-date as possible.

The measurements themselves have been compiled from British and American surveys, which I have compared with findings of my own and those of some of my colleagues.

General anatomical descriptive terms of position

The anatomical position

The anatomical terms of position assume that the body is standing upright and at 'attention' with the arms hanging by the sides with the palms of the hands facing to the front with forearms fully supinated. The feet face directly forward. From this position spatial relationships are communicated. (Figure 1a.) This is a conventional attitude known as the 'anatomical position', and not a functional posture. The normal posture is shown in the illustration immediately following. (Figure 1b.) The median plane divides the body into two equal halves, and is an imaginary perpendicular centre section from the top of the head to the base of the figure between the feet, at right angles

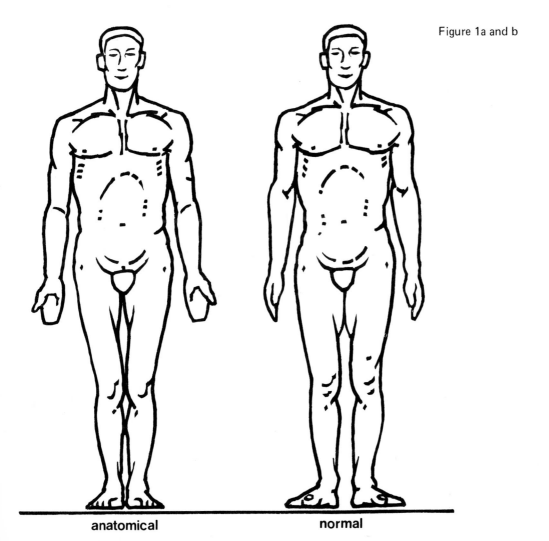

Figure 1a and b

anatomical   normal

to the direction of the shoulders. This plane is called the sagittal plane and any plane which is parallel to the median plane is a sagittal plane. Planes at right angles to the median and sagittal planes, and parallel to the direction of the shoulders, are called coronal. Completing the three dimensions of space, planes at right angles to both sagittal and coronal planes are called transverse.

Medial or *medial plane* means nearer, or a plane facing the median plane.

Lateral or *lateral plane* means farther away from, or a plane facing away from the median plane.

Anterior or *ventral* means facing the front, or to the front of a body part.

Posterior or *dorsal* means facing the rear, or to the rear of a body part.

Superior or *cranial* means in the upper portion or nearer to the head.

Inferior or *caudal* means in the lower portion or nearer to the feet.

Superficial describes a position on or nearer to the skin surface.

Deep describes a position nearer the centre of an anatomical part or farther away from the skin.

Distal and *proximal* indicate the relative ends of limb parts to one another, and they give positions in relation to the commencement of a limb. Thus the proximal end of the arm is the shoulder and the hand is at the distal end. However, by division down into smaller portions, the elbow becomes the proximal end of the forearm whilst the wrist is the distal end.

Anterior and posterior surfaces in the upper limb are also described as flexor and extensor. This terminology has to be completely reversed for the lower limb, posterior or dorsal surfaces becoming flexor and extensor being anterior or ventral.

All these terms compose pairs of relationship and used in combination enable an effective fix to be made on an anatomical site or position.

Weight of body parts

Table 1		Percentage weight of body parts
Head	from	6% to 8%
Trunk	from	40% to 46%
Legs	for both	30% to 36%
Feet	for both	3% to 4%
Arms	for both	10% to 12%
Hands	for both	1% to 2%

There will be variation in the weight of body parts due to the variation in physique and sex. The extremity of the head however is generally greater in weight than the extremities of the hands and feet added together. The proportionate weight of the head in relation to the rest of the body parts should be noted because quite often care for head weight by a head rest is a necessary ingredient in a human situation.

The 'normal' or functional position for the body

The body is standing upright and at attention, although the bracing back of the shoulders and lift of the chest is 'comfortable' and not exaggerated to the extent of a parade ground manner. The arms hang freely by the side, the lower arm being pronated in such a manner that the thumbs face anteriorly. The feet are positioned slightly apart with the medial surfaces of the feet making an angle of about 30° to 45° between them. The angle is rather smaller during walking.

The line for the centre of gravity of the standing figure in the 'normal' position can be imagined pulling down perpendicularly through the three component masses: the skull, the thorax and abdomen — pelvis. These body parts making up something like 60% or more of body weight. If this pull of gravitational force is examined in side view it will be seen that it must pass through the legs and be contained inside the arch of the foot. (Figure 2.) This is because, however varied man's structure may be, the centre of gravity must remain within its base, the feet. The normal standing position is in fact quite efficient; for the balancing trick that man is able to perform with the component parts of his body requires little muscular exertion. When standing restfully we relax our muscles but extend our legs and knees against the ground so that the ligaments of the joints brace them against collapse from the weight of the body above. The arches of the feet are retained by ligaments, although during lengthy standing foot and ankle joints will have to be reinforced by muscular lift and support.

The relationship of the centre of gravity to body parts, in side view, varies according to the figure type under examination. Its relationship to the head, the shoulder girdle and the vertebral column is very uncertain; at the hips it may either pass through the head or anatomical neck of the femur or at some point anterior to this. At the knees it will either pass through the patella or at points anterior or inferior to it, and it must then pass in front of the ankle joint through the arch of the foot. Different figure types have different postures, and there is not one normal posture to which all people can conform. Total height, leg height and body shape may be considered to be the most influential factors in posture.

Posture in young children is different, in as much as the knees are not braced back so firmly as in the adult, and the legs appear more bent.

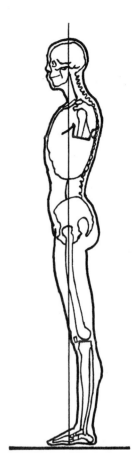

Figure 2
To show normal posture side view. Here the line of gravity passes down through the mastoid process on the skull, through the acetablum in the pelvis slightly in front of the knee and then passes in front of the ankle joint. It must be remembered that this line of gravity is illustrated on a figure type that is highly mesomorphic. Note the four good curves to the spine.

In more asthenic figure types the body parts would not appear to be so well balanced about a line of gravity, the head and cranial part of the trunk may be found to 'sway' in a dorsal direction.

1 Growth, maturity and old age

Growth

During the whole developmental period between infancy and maturity many demands are made on the growing child by its environment. Although we may consider that much of the child's response to new situations is genetically programmed, its continuing growth and ability to survive depends on a resilient and adaptable physiology working to sustain an ever greater mobility. This mobility is gradually attained by the extension and growth of the various organs and segments of the body. The growth rates of these organs and segments are not parallel but move a different speeds from each other.

A growing child exhibits a gradual change in appearance and shape. (Figure 3.) Superficially it can be seen that some parts of the body are growing faster than others. A good example of this can be readily seen in the head, where it can be observed that for many years during early childhood the face remains relatively small in relation to the rapid increase of size of the cranium. (Figure 4.) The brain, so necessary for initiating action, approaches adult size quite early. It reaches 80% of adult size between four and five years of age: the time for joining school. The sexual reproductive system waits until later, when sometime about or after eleven years it makes fairly rapid growth and the difference between the sexes becomes steadily more

Figure 3
As will be seen, human beings vary in proportions one from another. It is useful on occasions to have an average figure to refer to. In this illustration 'head lengths' have been used as a unit to obtain the varying bodily proportions found at different ages.

At one year a child is four heads high; at four years five heads, at nine years six heads; at sixteen years seven heads, and an adult is seven and a half heads. This is only a customary equivalent. A child at two years is half the height of an adult.

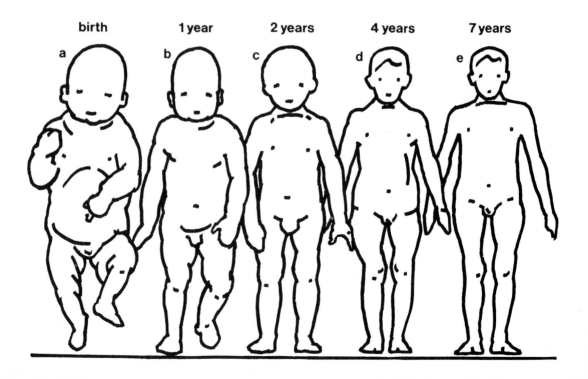

birth   1 year   2 years   4 years   7 years

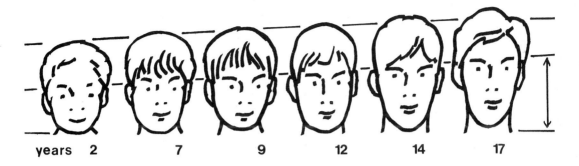

years 2   7   9   12   14   17

Figure 4
Comparative development of child's face and head

manifest. The two different speeds with which these body parts grow can be contrasted again with the increase in body weight. Body weight represents the sum total of all the body parts; and against time it can be informative about a child's progress towards maturity. Body weight shows a steady increase until about the sixth or seventh year of life when it is rather less than 50% of adult value. From seven to eleven years it almost marks time and so increases very little. (Table 7.)

During the 'marking time' phase (of weight increase) glandular changes take place which are in the nature of a prelude to the later stages of growth in the skeleton and body tissues. From about eleven years onwards body weight again shows a regular increase until maturity. This increase is represented by a steady gain in standing height or statue; which in the first place is the result of the increase of leg growth although arm maturity preceeds leg maturity.

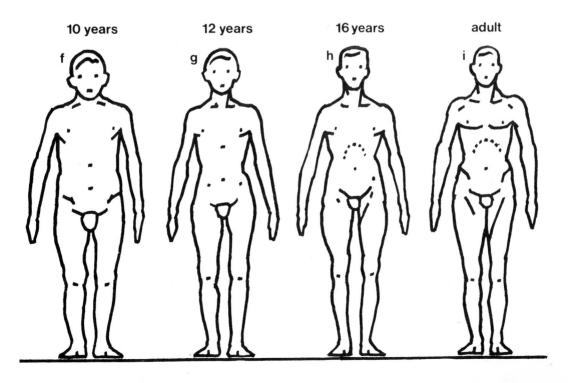

10 years   12 years   16 years   adult

It can be seen from these examples that there is often a wide variation of speed of growth between body parts, and always some variations while growth takes place. Externally this means that a child's body is continually changing its shape and therefore its proportions. So we can say that a child's body is liable and prone to change, and that these changes are normal and build up, stage by stage, towards a balanced mature condition.

The variation of the speed of growth between one body part and another is regulated so that each part reaches its size, proportional to the role it has to play in the body's physiology at the correct time. The endocrine glands provide the secretory centres which release hormones to stimulate the growth of particular tissues. One part of the brain, the hypothalamus, is the controller of these functions.

It is the alternation of the speed of growth between body parts that leads to a diversification and differentiation of one part of body structure from another. So it may be considered that the human body during development and growth changes in three ways: by increase in size, by differentiation of structure, and finally by the alteration of shape. These changes present three factors by which we may measure growth. The factors are three different types of dimensional increase: increase by length, increase by area and increase by weight. Betweeen birth and maturity the approximate increase overall in each of the three dimensions is: height, an increase of three and a half times: skin area, an increase of seven times: weight, an increase of twenty times.

Although these increases are dissimilar from one another

Figure 5
Three girls of 11.5 years

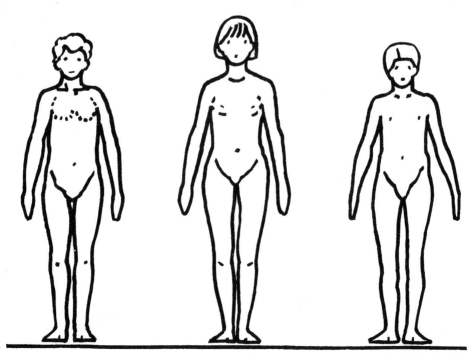

there is a typical mode of growth increment in each case.

It is because of the variety of the rates of progress in different dimensions during the growth of a child, that it is proper to consider it to be a process of development as well as growth.

The human being has the most protracted period of growth and development of any animal. Man's childhood, for instance, is about as long as the total life expectancy of a monkey. An average of 20 years is required for man to develop to full maturity. Twenty years is given here as a 'safe' figure for physical maturity. The numerical age of a child cannot be taken as a point of reference when establishing its progress towards physical maturity. To say that a girl is 13 is not any evidence of her having reached a particular stage in puberty. Nor is it any evidence of her size in any bodily dimension. For the designer numerical age is less important than body size and the proportions achieved.

When studying physical development, and leaving numerical age aside, we still meet difficulties. *With young children often the short ones grow more slowly; yet in adolescence the taller ones grow more slowly, and the shorter ones grow rapidly for a longer period of time. These type of changes in growth rates tend to accentuate the great variability of children's body measurements. For any numerical age, in either sex, we are likely to be presented with a wide range of body builds. (Figures 5 and 6 and Tables 2 and 3.)

*Tanner, J M *Growth at Adolescence*, Blackwell Scientific Publications, 1962.

Figure 6
Three boys of 12.5 years

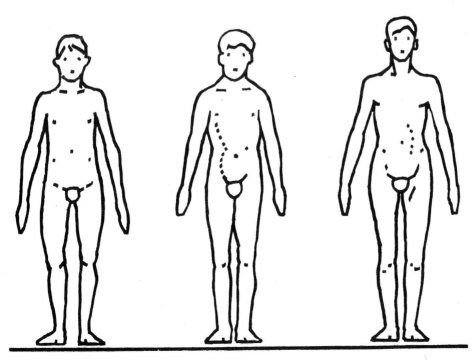

Table 2
Boys: Some typical length and girth measurement relationships found from age five to fifteen in millimetres and inches

Years	Arm length		Leg length (Crotch to ground)		Chest girth		Waist girth		Hip girth	
	mm	in.	mm	in.	mm	in.	mm	in.	mm	in.
5	444	17.5	483	19	533	21	508	20	546	21.5
6	470	18.5	508	20	546	21.5	508	20	559	22
7	508	20	546	21.5	559	22	521	20.5	584	23
8	533	21	572	22.5	559	22	521	20.5	584	23
9	559	22	610	24	610	24	546	21.5	635	25
10	584	23	635	25	635	25	559	22	660	26
11	610	24	660	26	648	25.5	584	23	686	27
12	648	25.5	686	27	673	26.5	597	23.5	711	28
13	673	26.5	724	28.5	698	27.5	610	24	749	29.5
14	698	27.5	762	30	737	29	635	25	787	31
15	737	29	787	31	762	30	660	26	826	32.5

Table 3
Girls: Some typical length and girth measurement relationships found from age five to fifteen in millimetres and inches

Years	Arm length		Leg length (Crotch to ground)		Chest girth		Waist girth		Hip girth	
	mm	in.	mm	in.	mm	in.	mm	in.	mm	in.
5	444	17.5	483	19	533	21	495	19.5	559	22
6	470	18.5	508	20	533	21	508	20	572	22.5
7	508	20	546	21.5	559	22	508	20	597	23.5
8	533	21	572	22.5	572	22.5	521	20.5	622	24.5
9	559	22	610	24	597	23.5	533	21	648	25.5
10	597	23.5	635	25	635	25	559	22	686	27
11	610	24	660	26	648	25.5	559	22	711	28
12	635	25	698	27.5	686	27	584	23	749	29.5
13	660	26	711	28	737	29	610	24	813	32
14	686	27	724	28.5	762	30	635	25	864	34
15	698	27.5	737	29	800	31.5	648	25.5	889	35

When considering the physical growth of the human body in detail we find that although there may be extension or growth in one segment in a transverse direction, in the immediate adjoining segment growth may be taking place in a lengthwise direction. Growth of body parts can be seen as an example of the biological law of alternation of growth. During growth the bones of the skeleton undergo changes of proportion both with and without the formation of cartilage growth-plates. (See Figure 9.) The law of alternation of growth can be seen to be clearly operating in the change of proportion taking place during the growth of the long bones of the skeleton. It can be considered that the long bones thicken for a period ranging from 4 to 6 months and then lengthen during the next 4 to 6 months. For example, the femur bone of the leg thickens while the tibia and fibula bones lengthen. Then in the next 4 to 6 month period the femur lengthens while the tibia and fibula thicken. Throughout the growing period there is an extensive remodelling of all the bones of the skeleton taking place in this manner. This remodelling is paced by growth in the superficial tissues, and as the internal structure undergoes some corresponding alteration takes place in the external forms of the body. A child's figure is not like an adult's in miniature. It re-forms and almost seems to deform, as it progresses. The same number of parts of an adult are present, but they are continually being presented in different proportions. As the law of alternation of growth suggests there are times when feet seem too big for legs; when hands seem to be too big for arms; or lower arms for upper arms; and these instances can be easily multiplied. (Figure 7.)

Figure 7
The consideration of the growth of a child's limb in segments which are growing alternately by length and by transverse section. This example is based on the possible development of a girl's arm

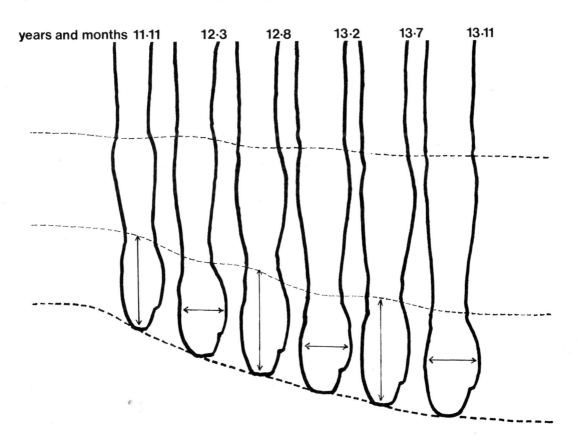

years and months   11·11   12·3   12·8   13·2   13·7   13·11

In early infancy there is a general chubbiness, with relatively large dimensions of the trunk and head. (Figures 3a and 3b.) The upper part of the body is proportionately larger than the lower. The cranium length and breadth and circumference are about 96% of the adult value by 10 years of age.

There is a great difference in the shape of the thorax between the new-born and the adult. In the first year of life the horizontal section of the thorax is cylindrical, the lower ribs are horizontal and the upper part appears constricted. (Figure 3b.) By the second year the horizontal section becomes oval. The vertical thoracic dimensions are shorter in infancy and the thorax appears very small compared with the abdomen. (Figure 3b.) The new-born are, in fact, all digestion or indigestion without means of locomotion.

Until about six years the limbs appear relatively short and puny. After six, during the period of juvenile growth, the legs and arms increase in relation to the trunk. (Figure 3e, f and g.) Throughout childhood the speed of growth for arms and legs remains different from the speed of growth for the trunk and head. With the growth of the limbs the growing child begins to fulfil the fuller role of the upright primate; maintaining its upright posture with greater ease and developing an ever freer use of the fore-limbs for grasping, swinging and manipulation, and also showing an ever-increasing tactile appreciation with the hands. At a later stage of development, with a larger chest and longer clavicles and greater power to pronate and supinate the fore-arms, the growing child can obtain a greater vigour of the arm and an increased dexterity of the hand. (Figure 3g and h.)

Until the end of the juvenile stage at about nine years the growth for both sexes is much the same.

During pre-adolescent years growth is largely concerned with building up a strong skeleton and extending tissues.

Between ten to eleven years, or the pre-puberal period, more height is continually being gained by the continuing growth of the lower limbs. (Figure 3f and g.) At about 12 years the cervical vertebrae lengthen and the neck becomes apparent for the first time. (Figure 3g.) The clavicles begin a descent to the horizontal which is their position in adult life. During this time the trunk retains a small appearance proportionately to the whole statue.

The waist, however, is becoming more defined and greater movement is possible between the pelvis and the thorax. At about this age the external physical differences between the sexes begin to make their appearance as a result of the processes of sexual morphological differentiation.

The period of puberty comes at different times for boys and girls; girls being something like two years ahead of boys. Girls are generally heavier, taller and have a greater body surface area than boys in the early pre-puberal period, but are then generally lighter than boys to puberty. Boys begin to assume extra weight in the later pre-puberal period.

The segments of the appendages continue growing, and in particular the proportion of leg length to trunk length increases through early and middle adolescence. (Figure 3h.) The thorax remains narrow, although the slope of the shoulders is beginning to increase because of the extra weight being gained in the arms.

From about 10½ to 11 years in girls and from about 12½ to 13 years in boys there occurs a remarkable acceleration of growth. This acceleration may be maintained for about 2 years in both sexes. (Figure 8a.) This acceleration is clearly noticeable visually, and is very strongly marked if data of one dimension of a growing child is plotted in graph form. (Figure 8b.) During the period of the adolescent growth spurt there is a very marked increase in size of various body parts. The spurt of growth starts with an increase in foot length. In general growth spurts commence in the distal segments of the appendicular skeleton and proceed through to more proximal ones. In the leg region, after the growth of foot length, the lower leg extends in length, and this is followed by more growth in the upper leg and thigh. At all ages of childhood the feet dimensions keep ahead of arm, leg and trunk dimensions and are nearer adult size. Increased spurts of growth in the feet may come a year to two years ahead of those of statue. Exceptionally rapid growth in the feet is exclusive to girls; and for those girls who have attained marked increase of dimensions in this region early, subsequently their feet will then cease to grow much more. Maximum velocity of foot growth can be reached at 10 years of age for girls, whereas

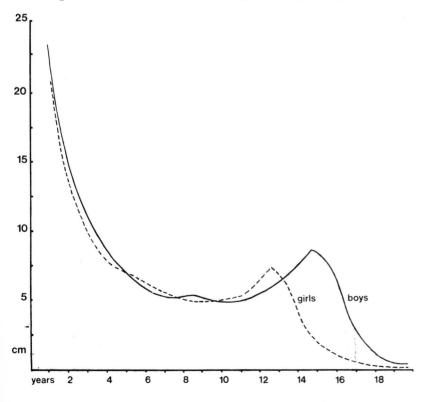

Figure 8a
Graph lines depicting comparative increases of girls' and boys' heights in centimetres each year. Upward peaks in teens are the adolescent growth spurt. The boys' growth spurt can be seen to be later, extending the pre-adolescent growing period and giving boys extra increments for standing height and also adding larger body dimensions generally

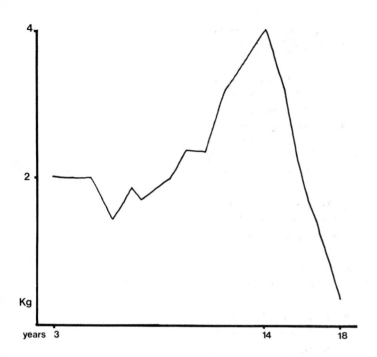

Figure 8b
Incremental increases in kilogrammes per year for a girl from three years to 18 years.

Maximum increase in this case was at 14 years. Child's final weight was about 49.89 kg (110 lb)

a boy's maximum is at about 12 or 13 years. Velocities of growth vary throughout post-natal development, and the increase of foot length growth has periods of acceleration. However, final foot growth probably takes place in breadth dimensions and not in its length.

Three to four months after the feet begin to increase the hip and chest breadths commence their growth spurt. Boys' shoulders and trunk height increase in relation to hip width and the definition of male proportions commences. (Figure 3h.)

Range of heights for each age group for either sex increases with age, although ranges of height do not appear quite so great between 6 and 12 years old. After 14 or 15 years of age until maturity the range of heights increases to an extreme of ± 203.2 mm (8 in.) each side of the mean value. 5th and 95th

Table 4
Variation in the range of boys' and girls' heights from birth to maturity. The range is from the 5th to the 95th percentile

	Range	*Difference*
Birth to 6 years	± 32 mm (1.25 in.)	64 mm (2.5 in.)
From 6 to 11 years	± 22.5 mm (.9 in.)	45 mm (1.8 in.)
From 11 to 14 years	± 32 mm (1.25 in.)	64 mm (2.5 in.)
From 14 years to maturity	From ± 32 mm (1.25 in.) at 14 years to ± 108 mm (4.25 in.) at maturity	64 mm (2.5 in.) to 216 mm (8.5 in.)

percentiles giving values of ± 101.6 mm (4 in.). The whole range of normal adult height can be found between 1397 mm (55 in.) and 2006.6 mm (79 in.).

Girls on the other hand have a spurt in hip-width growth and so begin to develop the typical female silhouette. Sexual differences of proportion become more marked in the trunk region with the development of breasts by the female. (Figure 10. Compare Tables 2 and 3.) Also with a less broad thorax and shoulder girdle, and a slightly lower sternum bone than boys, a girl's neck and shoulder regions begin to show their more characteristic delicate structure.

Leaving aside the development of breast tissue, boys and girls have a fairly similar increase rate for chest development. In the head the face has remained short during infancy but as the last of the permanent teeth erupt a spurt of growth takes place in the facial bones. The face becomes longer and the chin more prominent, and the more purposeful look of the adult makes its appearance.

In the arm region the sequence during the growth spurt follows a similar order to that of the leg. This is, the spurt of growth starts in the distal segment of the hand and by alternating four- to six-month periods progresses to the upper arm. The only distinction between the longitudinal growth of the skeletal parts of the arm and leg is, that whereas the growth of the bones of the arm takes place mainly at the proximal end of the humerus and the distal ends of the ulna and radius, this order is exactly reversed for the long bones of the leg; the femur bone growing mainly at its distal end, and the tibia and fibula growing at their proximal ends. It would appear that these main growing ends have been selected in the course of man's evolution, so that the joints of the elbow, hips and ankle are prepared early in a child's development to withstand the stress and strain of his erect position and the practice of his physical functions. (Figures 9b and c.)

In early and middle adolescence it is the leg growth that takes the entire body towards mature height. Mature proportion and final height is obtained in late adolescence and sub-adulthood by growth of the trunk. Leg height and trunk and head height are the chief constituents of stature. (See Table 5 for typical heights.)

Growth does not cease abruptly, but tails off finally in the regions of the trunk and the head. The trunk may continue to increase in height until 25 years of age or more, and the head measurements may continue to increase, in a very small percentage of cases, for a much longer period. Forms of late growth can be confusing as there is a possibility that one may be observing pathological over-growth. These are restricted individual concerns and lie outside the designer's appraisal of growth.

We find the most obvious sex differences in body physique illustrated by changes of body size and shape. Differential

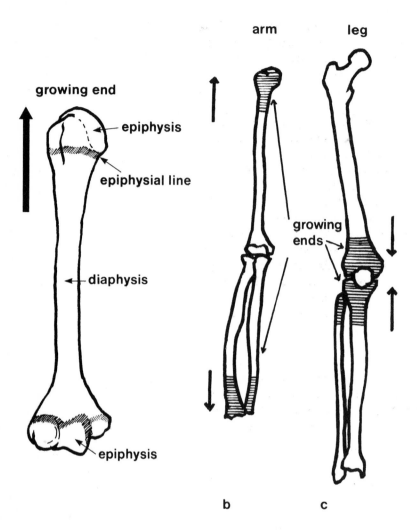

arm **leg**

growing end

epiphysis

epiphysial line

diaphysis

epiphysis

growing ends

b c

Figure 9
Growing ends
Growth continues for the greatest time at a preferred end of a long bone. The growing ends of the arm and leg bones are illustrated here in **b** and **c**. **a** illustrates diagramatically the portions of a growing bone. The central section or primary centre is the diaphysis and the secondary centres at the ends are called the epiphysis. An epiphyseal section of cartilage separates the two centres and it is its constant adaptation to the processes of ossification taking place on its borders that facilitates bone-length growth. Finally, at about the time of puberty the primary and secondary centres of ossification unite and the proportions of the adult bone begin to be laid down

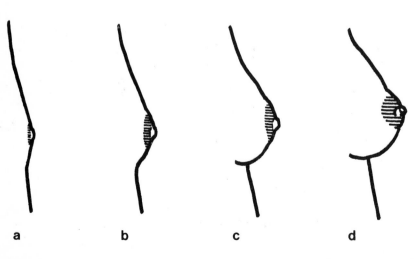

a b c d

Figure 10
Breast development
a Breast bud. The commencement of the body of the breast

b The enlargement of the breast body and the definition of the areolae as separate eminences

c Continuing enlargement of the breast body and the separate and distinct elevation of both the areolae and the papillae

d Adult stage. The attainment of the conical or hemispherical form of the mature breast body with the papillae fully elevated

In the diagrams the areolae are indicated by shaded areas

growth rates at adolescence in the sexes vary as regards shoulder and hip breadths, trunk heights, head size and the proportions of arm and leg segments. (Figure 11.) The adolescent growth spurt occurs later in the male than the female. The overall difference in size and proportion between adult men and women is largely the result of the delay in the onset of the adolescent growth spurt in boys. This longer growth period for men gives them larger bones and a greater body size. It also adds extra increments to standing height. (Figure 8a.) Boys from birth are averagely bigger at all stages of growth except during the early part of the adolescent growth spurt.

Boys' and girls' strength and endurance increase with the growth of bone and muscle. Generally the peak of strength comes about a year after the achievement of the peak of height.

Although boys wait until a later age than girls to gain maximum strength it is then superior to girls' due to the larger muscles and bones they have built up. Greater strength in boys and men is also the result of the broader shoulders of their sex, which is a noticeable feature of male sexual dimorphism as the broader pelvis is a feature of the female sex.

Maturity

In general the differences between the sexes in the skeleton are as follows. All the eminences of the male skull are more angular and nearer to 90° than the female; the only exception being the temporal eminences, which are more marked with the female than with the male.

Table 5
Typical heights for both sexes from 1 to 19 years of age in millimetres and inches

	Male mm	Male in.	Female mm	Female in.		Male mm	Male in.	Female mm	Female in.
Birth	505.5	19.9	503	19.8	11 years	1442	56.4	1455.5	57.3
1 year	752	29.6	744.5	29.3	12 years	1500	59.3	1514	59.6
2 years	876	34.5	871.5	34.3	13 years	1561	61.5	1556.5	61.3
3 years	960	37.8	952	37.5	14 years	1647	64.8	1592.5	62.7
4 years	1033.5	40.7	1031	40.6	15 years	1717	67.5	1613	63.5
5 years	1109.5	43.7	1109.5	43.7	16 years	1753	69.0	1620	63.8
6 years	1168	46	1156	45.5	17 years	1765	69.5	1622.5	63.9
7 years	1224	48.2	1219	48	18 years	1770	69.7	1626	64.0
8 years	1283	50.5	1226.5	50.3	19 years	1774	69.8	1628.5	64.1
9 years	1336	52.6	1328.6	52.3					
10 years	1384	54.5	1389.5	54.7					

At 2 years approximately half of adult height has been achieved.

12 years   **13 years**

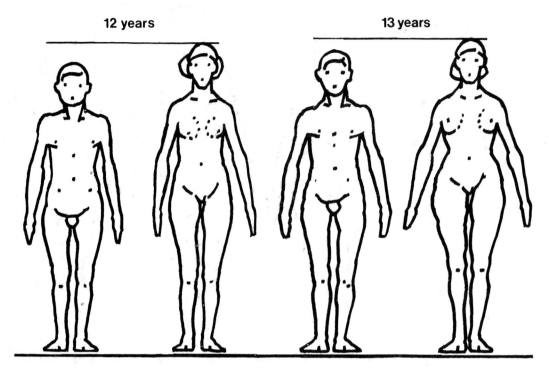

The male trunk shows a broader and longer thorax with longer clavicles and greater scapulae. The male neck appears shorter because of the higher position of the manubrium in relation to the vertebrae prominens. The female pelvis is wider and altogether more spacious. The sacrum is much nearer the horizontal, the inner surface facing downwards, and there is often a smaller lumbar-sacral angle. This gives the female a more mesomorphic appearance at the pelvis in side view. (Figure 20.) The area of the ilium is less but the pubic arch is lower and wider, and the acetabulum face more anteriorly and inferiorly.

Figure 11

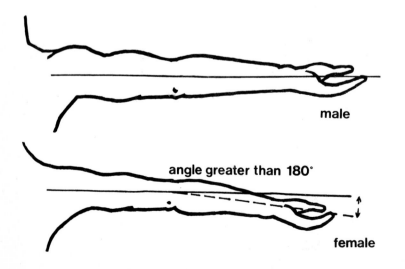

male

angle greater than 180°

female

Figure 12
Elbow extension, anterior view

Male maximum 180°

Female: a percentage can manage to produce an angle greater than an 180° with this movement

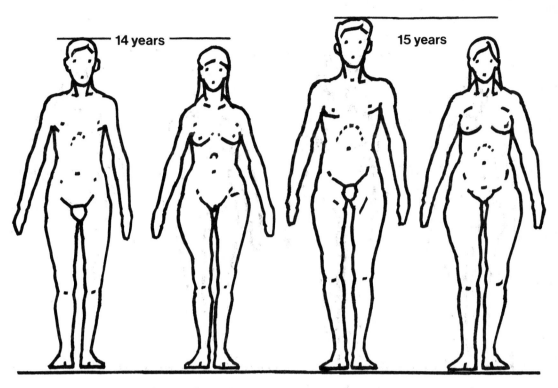

14 years — 15 years

The greater width of the base of the female pelvis means that their pubic-femur relationship is nearer an equilateral triangle (Figure 13.) The bones of the legs and feet of the male are longer than the female. The bones of the arms and hands are also longer, but this may be less visually obvious than the difference in the legs. 85% of females can extend the elbow joint to beyond 180°, and have a proportionately longer humerus bone in relation to the radius and ulna bones. (Figure 12.)

Some of the more obvious differences in the skeleton between the sexes may be summed up as follows: the male has

Figure 11

Table 6
Possible decline in stature in millimetres and inches

| Years | Male | | Female | |
	mm	in.	mm	in.
20—24	None	None	None	None
25—29	None	None	5	.2
30—34	5	.2	2.5	.1
35—39	5	.2	2.5	.1
40—49	10	.4	5	.2
50—59	12.5	.5	10	.4
60—69	15	.6	15	.6
70—79	7.5	.3	15	.6
80—89	7.5	.3	None	None

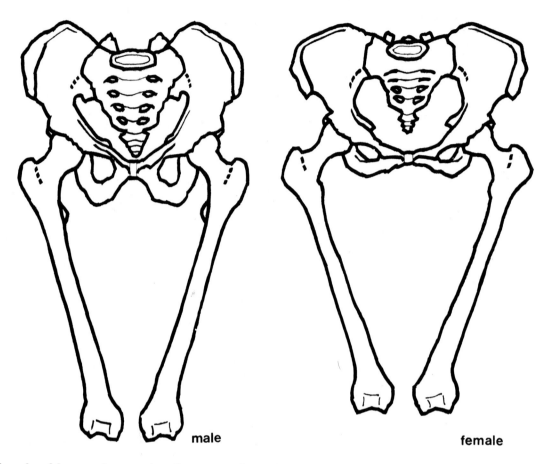

male **female**

Figure 13
Femur-pelvic relationship

wider shoulders and a greater thorax, and a relatively narrow
pelvis; a larger skull and longer arms and legs with bigger hands
and feet. The female has narrow shoulders and a rounder and
slighter thorax with a relatively wider pelvis which is tilted more
anteriorly; a smaller and more rounded skull, shorter arms and
legs with smaller hands and feet. The average difference between
men and women's heights is about 6 or 7%.

The final picture of the male and female form is made up by
differences between superficial body tissues. The male figure
has more muscle to fat than the female. The average male will
have a ratio of six to three of muscle to fat; the average female,
a ratio of five to four of muscle to fat.

From these ratios it is at once evident that there is a predominance of body fat in the female. Girls also have more
subcutaneous fat than boys from birth, more noticeably so
from about seven years onwards.

With the rapid expansion of the transverse sections of the
body segments during the adolescent growth spurt, girls' growth
is more in fat than muscle, though it may be a less apparent
accumulation in the arms. In the trunk and upper thighs the
acquisition of fat by growing girls can be quite large. Boys show
varying amounts of body fat up to adolescence, and then with

Table 7

Typical weights for children, men and women

| | Boys & Men | | Girls & Women | |
	kg	lb	kg	lb
2 years	12.70	28	12.25	27
4 years	16.33	36	15.88	35
6 years	20.42	45	18.60	41
8 years	24.95	55	24.95	55
10 years	31.75	70	31.75	70
12 years	38.10	84	39.92	88
15 years	56.25	124	52.16	115
19 to 20 years	70.76	156	55.79	123
20 to 30 years	73.48	162	58.97	130
31 to 40 years	75.30	166	60.78	134
41 to 50 years	76.66	169	65.77	145
51 to 60 years	75.30	166	67.59	149
61 to 70 years	73.94	163	66.68	147
71 to 80 years	68.95	152	63.50	140

Weights

Only 10% are over 90.72 kg (200 lb) in weight.

Typical women's weights may range from 5th percentile value of 45.36 kg (100 lb) to 95th percentile of 79.83 kg (176 lb).

Typical men's weights may range from 5th percentile value of 57.15 kg (126 lb) to 95th percentile of 96.61 kg (213 lb).

an increase of bone and muscle growth to fat, give the appearance of the 'lean and hungry youth'. A boy's growth spurt is in bone and muscle, when his large and more powerful physique is being assumed. Boys growing in muscle and remaining stationary in fat will actually appear to loose fat. After the adolescent growth spurt a proportion of boys will assume extra fat again as their bodily constitution consolidates its mature form.

The manifestation of particular ratios of bone to muscle, and muscle to fat during adolescence is the determinant of sex in physique.

Subcutaneous fat in the normal adult has a predilection for residence in particular body areas. This is so for both sexes, although its flowering abundance in certain areas is reserved for the female during sub-adulthood and maturity. The reason for the residual areas of subcutaneous fat on the body is not altogether clear. In particular regions it helps to conserve body heat; it does not remain inert however and is continually being mobilized for metabolic breakdown. Perhaps it is a remaining stage from man's earlier history when it was necessary to carry one's own emergency rations. The female role of nursing mother

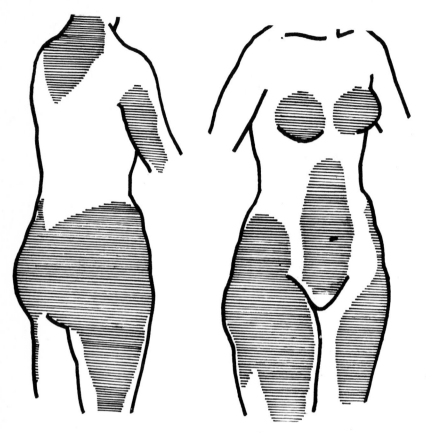

Figure 14
Resdidual areas for body fat

would make this doubly necessary. The residual areas for body fat are: the buttocks and the flanks; the abdomen down to the pubic arch; the anterior and lateral surfaces of the upper thigh; the breasts or pectoral region; the back of the neck about the vertebrae prominens, and the posterior of the upper arm. (Figure 14.)

The pectoral region is a site of fat tissue which contains the mammary glands. In the female the mammary glands, or breasts, grow and elaborate during and after puberty and the fat tissue is also increased. The breast is at first disk-like, and next assumes a conical appearance and finally a hemispherical form. (Figure 10.) The breast tissue formation is without any structural support; because of this fact and the infinite possible range of breast proportions it is impossible to place much reliance on transverse circumference measurements taken at nipple section.

It can be seen from examining the list of residual areas of fat that the greater part of its accumulation is situated about the thighs and hips. Accumulations of fat are therefore more readily noticeable in this region, and they are a typical female characteristic. The fat of this region has been found to have a higher correlation to body weight than the fat in any other region, on female adults. The heavier women will therefore always have the largest hips. Correlations of the fat of other body regions to weight are very much lower and have no reliability. In the more prosperous populations body fat increases in both sexes with age; the middle and early old age groups showing tendencies towards corpulence.

There is evidence of the effect of a superior geographical or social environment on total body growth. There have been indisputable increases in growth rates in the more prosperous countries, and there is a definite trend towards earlier maturation and greater total body build. This does not mean that a definite increase of nutrients will always produce a definite amount of body tissue.[1] Genetic and environmental factors combine to inject a series of unknowns into growth rates of children from a similar population; equally well nourished or not.

There are regional differences in growth that cannot be explained in terms of nutrition or climate, but would seem to depend on genetic factors peculiar to the region.[2] There can be some delay in growth rates of children in hot climates.[3] These differences may be in part responsible for the range of physique to be seen in man globally. The findings of anthropometric surveys in England of children at adolescence show that a superior environment can produce an increase in total height from 38 mm (1½ in.) to 50 mm (2 in.). An inferior environment might cause a loss of expected height of 38 mm (1½ in.). During infancy loss or gain shown in total height, due to the same factors, is about 13 mm (½ in.). There is evidence that suggests these figures would be true for European and North American populations.

Because of hereditary and environmental influences we can find great variability in children's growth rates. Identical children, other than identical twins, do not exist. Average body measurements for children cannot be characterised by a mean value obtained against age. Numerical age has no great significance when evaluating children's sizes for design purposes. The two quantities needed are body dimensions and their velocities of growth. (Figure 15a.) A series of measures of one particular body dimension from a child, taken over a period of years would show incremental increase from which the velocity of growth of that dimension could be determined. A series of these measures obtained from a number of children, and then stated in graph form, and the curves matched, so that maximum and minimum velocities coincided irrespective of age, would allow an average curve or parameter to be drawn. (Figure 15b.).

The examination of average curves of children's body measures would enable a designer to determine where positive

[1] Tanner, J M, *Human Biology*, Part IV, Factors affecting growth, page 343.

[2] Tanner, J M, *Human Biology*, Part IV, Regional differences of growth, pages 345 to 347.

[3] Weiner, J S, *Human Biology*, Part V, Climate and growth, pages 457 to 459.

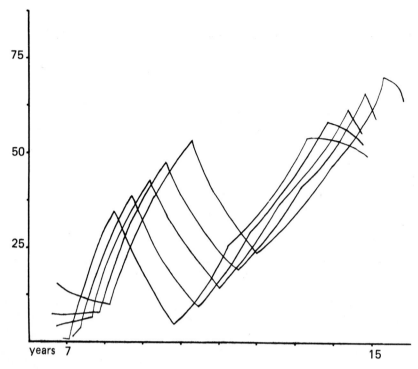

Figure 15a
This shows the manner of increase of arm length for five boys from 5 to 16 years of age. The amount of total increase is similar through the range for all the boys, that is approximately 279 mm (11 in.) The difference of overall length before the growth noted here took place was 102 mm (4 in.). The smallest boy having an arm length of 394 mm (15.5 in.) approximately, the longest an arm length of 495 mm (19.5 in.) approximately

changes in terms of increases in dimension make a new size necessary to be fitted. He would as well have a record of the change of body dimensions. This method would be equally suitable in any sphere of children's requirements where a minimum or maximum range of design sizes needed to be calculated.

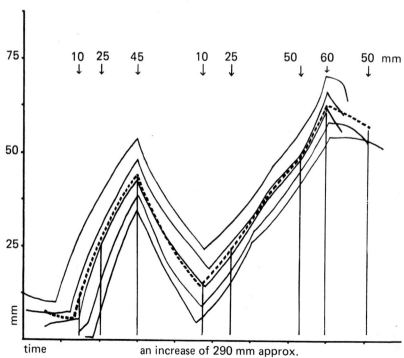

time   an increase of 290 mm approx.

Figure 15b
The average curve here simplifies the estimation of the increase in the boys' arm length. The five curves have been moved so that their peaks and troughs more or less coincide. It can now be seen that the mode of increase for all the boys is similar although each one took place at a different age range.

The incremental additions chosen on this example are marked by perpendicular lines. The amount of the increment is given above the perpendicular line in each case. Other incremental points might have been chosen. Those chosen would depend on the experience of the designer concerned. For example a clothing designer would be able to select the number of sleeve sizes that were necessary to cover the mode of increase of arm length. Other dimensions of a growing child's figure could be studied in a similar manner

Ageing

The process of ageing is a gradual one and it is impossible to determine with any accuracy an average chronological progression of features which would be of much use. Old age is an easily recognisable phenomenon, but difficult to assess in terms of loss of potential or useful value.

Examination of bodily tissues shows that the ageing process is a gradual series of degenerative changes in the nature of morphological systems; like the processes of growth, the processes of degeneration in body parts do not necessarily proceed at similar speeds. Some morbid changes in structure precede others. One of the changes is the diminishing power of cartilage to maintain its elasticity. Calcification of cartilage becomes more pronounced with advancing age. Another is the increased fragility of the bony skeleton. The remodelling process whereby bone renews itself gradually ceases; bones become slighter and more brittle with age. The nervous system seems better able to maintain itself. Of all the different types of cell in the body those of the nervous and skeletal muscular systems have the longest existence. Nevertheless muscle atrophy is a well known symptom of old age. On the other hand the velocity of nervous conduction diminishes very little in the ageing person and so one can speculate that for a moderate period it may compensate for a less adequate mechanical system. However after about 65 years of age there is a recognisable loss of motor functions, and a decreasing sensitivity to external stimuli. Activity is gradually reduced and there is a period of natural decline of bodily function. Degenerative out-growths occur from the margins of bones, which together with other age changes in joint tissue, and the history of traumatic effects, makes movement more difficult. There is an increased incidence, with age, of a variety of forms of rheumatism, and other diseases of a crippling nature.

The respiratory process and heart pump action also diminish in efficiency. All of these occurrences further restrict or slow bodily movement.

Outwardly old age shows itself in the face in a reversal to infantile proportions; mostly due to distortion by the over-growth and decay of the teeth, and malformations and wear in the jaws. The skin surface generally appears to wrinkle and sag and becomes coarser in texture on some parts of the body. On the face and backs of the hands the skin decreases in thickness, but it increases in thickness over the thoracic spine, the heels and in the post-deltoid region. Fat may cause loose and unsightly folds. Veins are noticeably larger. Due to calcification or atrophy of intervertebral discs there is a loss of movement in the trunk as well as a loss of height, but this last point can be overstated. (Table 6.) There are changes in posture as a result of weakening musculature and joint distortions. (Figure 16.) The stooping look of advanced old age is accentuated by gross

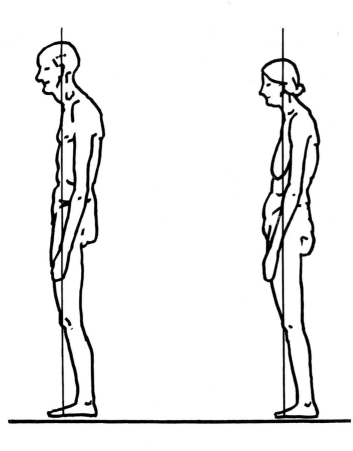

Figure 16
Posture of old age shown against
the line of gravity

changes in the lower limbs and region of the hips. In the earlier stages of old age body weight often increases and this throws more strain on a weakening musculature and the legs, as well as on the heart functions.

Maintaining a working posture becomes increasingly difficult for elderly persons and they require a slightly higher work surface for comfort. They dislike bending.

Ageing has an insidious onset and in one sense it can be said to have commenced at any time after the peak of physical activity has finally passed in the late twenties and early thirties. Thereafter there is a very gradual decline of bodily powers, which is a part of a long chain of physiological events from birth to death. In general there is a 50% loss of strength between 30 and 70 years of age. Without the intervention of pathological causes the percentage loss of strength will not be very large until after 40 or 45 years of age. The percentage loss between 30 and 35 years of age in strength is small; and in terms of exact measurement it is extremely difficult to show the slight physical limitations for a particular person in this age range. In a normal healthy person, used to an active situation, it may still be extremely difficult to detect significant loss of physical powers up until the middle or late fifties.

Marked physical incapacities are easily recognised. But the

slighter restrictions on human activity and their effect on performance are hidden from view by the human desire to exert maximum force and effort whatever the cost. We find with human performance that percentage rate of performance depends on the percentage rate of success achieved depending again on the limits of activity particular nervous systems can tolerate.[1] Success often comes by a combination of a process of trial and error and a good memory. The older age groups can be rich in these qualities which we call experience and wisdom. Experience and wisdom find means of circumventing disabilities and restrictions, obscuring known physical limitations, and achieving success. Any fruitful investigation of declining bodily mechanisms and powers would have to be an association of the psychology of ageing with physiological and anatomical studies. Meanwhile we can say that the wisdom and intelligence gained through a working life can prolong its usefulness.

[1] Thompson, d'Arcy, Ed Bonner, J T, *Growth and Form*, Cambridge University Press, 1962.

2 Figure typing

All human populations display distributions of varying physiques. Man is a polymorphic species, an animal which derives advantages from propogating a range of body or figure types. One obvious example of this is the marked difference in shape and proportions between the sexes; further proportional variations may be noted when a population from one geographical region is compared with another, of a similiar sex, from an entirely different world region. Finally it can be observed that each adult body form is built up by a number of morphological units each one of which varies from individual to individual of the same sex. Small differences in body proportions, almost impossible to detect at birth, are continuously multiplied throughout physical development and finally the individual morphogenotype or figure type is displayed.

The general aim of an investigation into the range of figure types in a population would usually be to attempt to improve the classification by the analysis of significant physical characteristics. Much present analysis uses multi-variate statistical methods, but there remains a constant need for schemes of visual inspection.

Most of the physical characteristics are quantitative like those of length of body parts, girth dimensions or weight. Some are qualitative like those of type of body hair or skin colour. It is the field of quantitative characteristics that is the special concern of anthropometry for designers or for the applied physical anthropologist. Quantitative characteristics include all external measurements of human body dimensions. Other quantitative characteristics that can be measured include angular and linear ranges of body segments and postural and limb forces and strengths. These two sets of quantitative characteristics of body measurements are sometimes called static and dynamic anthropometry.

When we examine a range of human physiques we are left with an impression of varying degrees of muscularity, of fat or corpulence, of small wrists or long feet and so on. It can be seen that to discover a satisfactory relationship between body parts is a formidable task. Neither is it easy to visualise overall differences in shape, which includes posture, between any two individuals from only a knowledge of their heights, any other length measurements, girth measurements and their weights. It is because of this difficulty that the need for the classification of body shape has arisen. Figure typing systems attempt to sum up as simply as possible the body shape variability in a population.

During this century and the end of last anthropometric surveys and field work have collected a great deal of metrical information about differences in body physique and constitution. Although precise measurement of physique is relatively modern, since the beginning of civilisation there has been a curiosity and a cultivated interest on man's part about his own shape. The progress of this understanding has intermingled with man's quest to understand his own substance and its history.

One of the most important aspects of this understanding has been the development of human biology as a separate discipline which includes taxonomy, polymorphism, work capacity, as well as the classification of physique, in its studies.

In the development of diagnostic practice from the time of the Greek fifth-century medical writer Hippocrates of Cos a number of attempts have been made to classify the inter-relationship of elements that determine a figure type. An important part of the examination of body constitution has always shown concern for enquiry about body shape. A number of attempts have been made to classify body shapes into types which share common normal or pathological features. Outside the medical field these classifications have been of continued interest and benefit to anthropologists and designers. Leonardo da Vinci attempted to correlate body segment proportions by geometric methods, showed interest in biometry and generally assisted the advancement of anatomical studies.

Hippocrates determined two polarities in the range of man's figure types: the phthisic habitus or the thin and tall physique, and the apoplectic habitus or the stout and short physique. These two polarities bring us surprisingly near present conceptions of a fat component running into a thin component. There is however no mention of a muscle or bone component. Maybe this third component was implicit in the Greek idea of 'wholeness between two extreme poles'.

Notable developments in figure typing theories were made by an Italian physician, Viola, and a French physician Rostan, and by a German psychiatrist, Kretschmer who each in turn advanced our conception of physical consitution.

After Hippocrates subsequent methods of figure typing produced formulas that stated all physiques in any population could be divided discretely and placed into a number of catergories. Viola however acknowledged the presence of a 'mixed-type' of physique that contained some of the elements found in the three primary groupings he had devised.

His three primary groupings also indicate his objective concern for the mixture of component features seen in any single physique. He diagnosed a Longitype with long limbs and a wide thorax, and a Brachitype with exactly opposite proportions: the third type, the Normotype contained all those physiques shading between the first two types. The fourth, and most significant a 'mixed-type' contained all the physiques whose anomalies excluded them from a place in the first three groupings. The 'mixed-type' is rather more the rule than the exception in the study of physique. With the conception of the 'mixed-type' there was the idea, current for the first time, that there can be anomalies occurring in some of the components of a physique which nevertheless do not place them in an abnormal category. In 1940 William Sheldon, an American psychologist, after much patient research and brilliant analytical thinking, devised what might well become the classical method of body-typing. Although doubts have been cast on some aspects of Sheldon's method the feasibility and usefulness of it has been proved, in particular by Dr J M Tanner who has refined and advanced the use of the system.*

Sheldon's system took the concept of the 'mixed type' of physique further in that he stated that any physique contained three components, but in a varying ratio for each individual. These variables are continuous in any population; and are the expression of its diversity.

William Sheldon's research was carried out at Harvard University prior to 1940, when his findings were published. His specimen population consisted of 4000 students. He used a combined method of photography and anthropometry.

Each subject was photographed in side, front and back view.[1] The photographs were carefully sorted and by visual comparison he found three extreme forms of body shape. These three extremes were reasoned to be the three aspects or factors of the total morphological structure of any individual. Each extreme was, in fact, a record of the end of the range of a component in physique maximised in one figure type. The range of each component was traced through the photographs by constant comparison and re-shuffling. The majority of the subjects did not show extreme forms of the components but were rather a more moderate blend of them. The three aspects or components of bodily morphology he had determined were called endomorphy, mesomorphy and ectomorphy. The three descriptions were derived from the terms used to describe the three initial layers in the early embryonic forms of higher life.

Endomorphy, as a component, describes predominance superficially in fat giving a roundness and fullness of shape. In deep morphology it indicates a large digestive system. With a round head and weak appendages the endomorph retains something of an infantile appearance.

Mesomorphy, as a component, describes a predominance in muscle and bone, giving an angular and hard shape. In the morphology generally it indicates a strong bony frame and connective tissue. With little subcutaneous fat and a resolute stance, the mesomorph is very much the ideal male and the person of action.

---

[1] For this kind of photographic work the subject has to be posed in a careful manner, and the working distance between the camera and subject so arranged that the optical problem of parallax error is avoided.

Ectomorphy, as a component, describes a predominance of skin surface relative to body mass; ectomorphy in extreme exhibits long thin body segments giving a weak and poorly balanced posture. In deep morphology it indicates a large central nervous system and brain. The thin appendicular skeleton by contrast makes the trunk look too small and the head too large, and there is a consequent suggestion of immaturity in this body shape. A list of the predominant differences between endomorphy, mesomorphy and ectomorphy are listed in Table 8.

Table 8   Some descriptive features of Sheldon's three components

---

Endomorphy male and female   (Figures 17 and 18.)
Total physique. Round and soft with large fat storage. In extreme form pear shaped. Abdomen is full and extensive and the thorax appears small. The limbs appear short and ineffective. Shoulders are full and round supporting a rounded head. The skin is soft and smooth. Fine hair, with little showing on the body.

Endomorphic skeleton
In side view the vertebral column appears straightened in the thoracic region. Bony girdles of trunk and pelvis are approaching the circular. All the bones are small with tuberosities and projections rather rounded.

Head and trunk
Features small and unobtrusive. Small cranium in relation to wide palate and face and a short neck. The trunk is long and heavy at the base. The greatest body breadth is near the waist which

Figure 17
Extreme endomorphic components

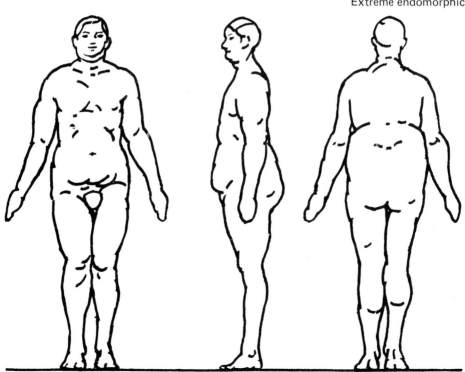

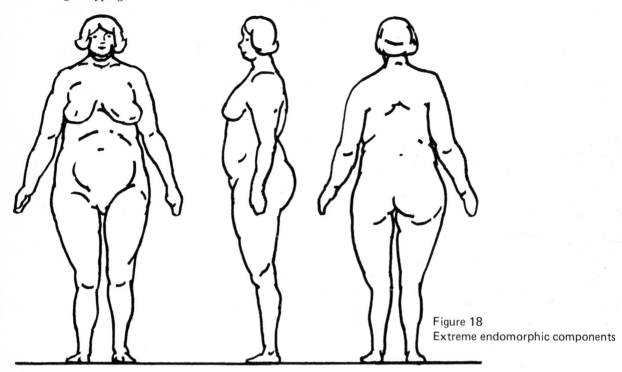

Figure 18
Extreme endomorphic components

Table 8 continued

is difficult to determine. The thorax is wide at the base. The whole trunk and head has a forward pushing penguin-like appearance.

Arms and legs

The limbs are short and from an inflated appearance proximally they rapidly taper down to small and weak distal extremities. There is no angularity, and so in the male where there can be the appearance of breasts, there is a suggestion of femininity.

Mesomorphy male and female   (Figures 19 and 20.)

Total physique. Square and rigorous in appearance with much prominent muscle. Shoulders predominate with thorax wide at the apex, and the abdomen small. The limbs appear large and strong. Neck is strong with a rugged head showing prominent eminences. Skin is rather coarse, as is the hair.

Mesomorphic skeleton

All the curves of the vertebral column are well shown. Bony girdles of trunk and pelvis are wider laterally than in the anterio-posterior dimensions. All the bones are heavy, with all the tuberosities, ridges and processes well defined.

Head and trunk

The face is large compared with the cranium. The forehead is often shallow and the facial eminences and angularities tend to approach $90°$. The neck is long and the sterno-mastoid triangle well marked. The thorax is massive, and certainly in a man, markedly broader than the pelvis in the region of the xiphoid process. The waist is low with the abdomen almost perpendicular in profile.

Arms and legs

These are generally heavy and well muscled, particularly at their proximal ends. The hands and feet are also large and capable in appearance. Often the forearm and calf muscles are very prominent.

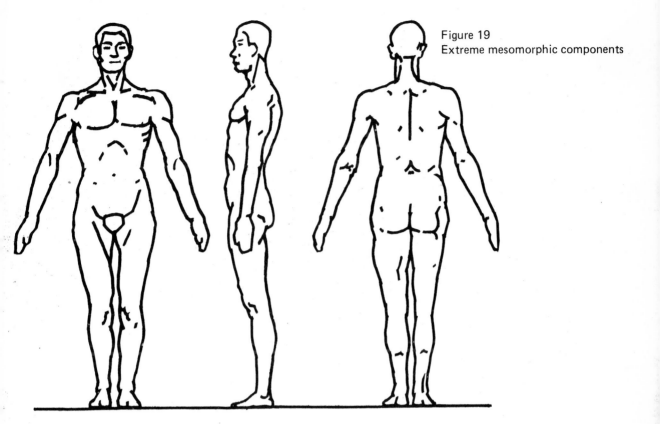

Figure 19
Extreme mesomorphic components

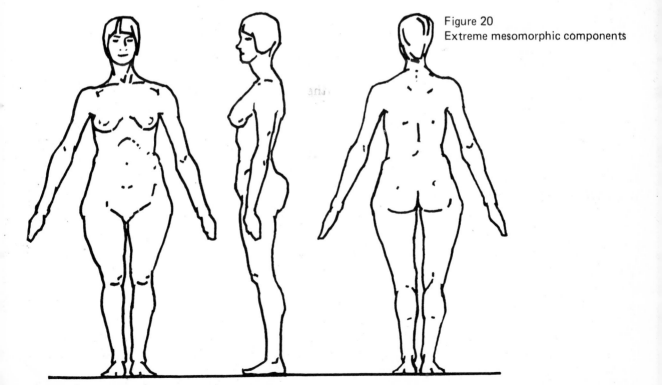

Figure 20
Extreme mesomorphic components

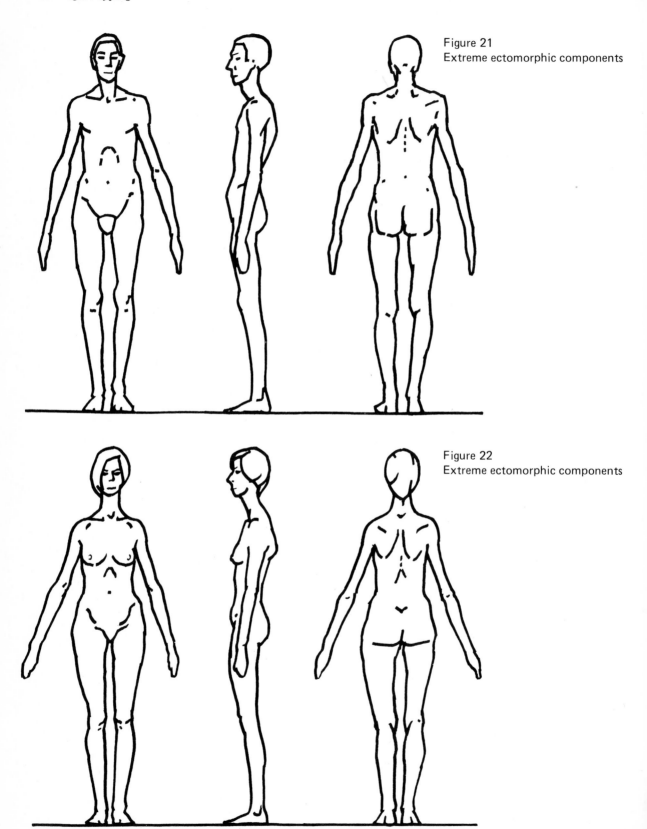

Figure 21
Extreme ectomorphic components

Figure 22
Extreme ectomorphic components

Table 8 continued
Ectomorphy male and female   (Figures 21 and 22.)
Total physique. Fragile and slender with a minimum of either fat or muscle. The trunk generally appears short and poorly positioned and is accompanied by long spindly limbs. Shoulders are wide but droop, the neck is slender and can appear inadequate for the head, which often has a large cranium. The skin is thin, hair is brittle and very variable in amount.

Ectomorphic skeleton
In the vertebral column the cervical and thoracic curves are well marked to the point of distortion, whilst the lumbar curve is flattened. The sub-costal angle is acute although the thorax is shallow in anterior posterior dimension particularly in the region of the sternum bone. Bones are light, but also variable in length, for stature in this component has a wide range.

Head and trunk
Features are finely etched with a slight chin. Brow can be large, the cranium and brain very extensive. The thyroid gland is prominent, and the neck hangs forward from rounded shoulders. The trunk is shallow and flat, the abdomen protruding in front of the thorax in profile. The distinct sag forward of the abdomen is due to ineffective musculature.

Arms and legs
They appear weak and long with a slight musculature. Their weakness is more noticeable at the proximal ends; and distally the toes and fingers are long and delicate.

The connection between Sheldon's components and embryonic layers may be tenuous and beyond proof, yet they do give the best means available for evaluating body shape. Any whole physique, or any anatomical segment of a physique, can be represented in terms of a combination of three components. A short training would enable a designer to think in terms of these components and develop an awareness of the nature of the diversity of human physique.

Sheldon assigned a scale of numerals for each of his three components. By anthropometric method[1] and visual photographic inspection he established a scale of numerals from one to seven in each component. The scale of numerals was equally spaced and became ratings in the system which he called somatotyping. The scale of numerals shows the strength of any one of the three components in any one physique. A somatotype rating indicates shape alone, not size. Two individuals of similar build may have the same somatotype rating although one may be larger than the other.

Somatotype ratings give the component of endomorphy first, then that of mesomorphy and finally ectomorphy. So a somatotype rating of extreme endomorphy would appear as 7—1—1; a somatotype rating of extreme mesomorphy 1—7—1; and a somatotype rating of extreme ectomorphy 1—1—7. These extreme somatotypes are quite rare. There are only approximately 15 present in any population unit of a thousand persons. More common ratings present in any population are

[1] Each fresh somatotype when rated was compared with other somatotypes whose subjects shared a similar height to cube root of weight ratio.

4—4—3 or 3—4—3, ratings 2—6—3 and 4—6—2 being more unusual.

Table 9 gives the approximate incidence of a variety of somatotypes in a population of Europids. Sheldon's system can be used in any geographic region, and according to the region the ratings would only differ in terms of their percentage distribution.

Looking at Table 9 it can be seen that in the more common ratings endomorphic and ectomorphic components never reach 6, and the mesomorphic component does so only twice. Over half of the physiques are dominated by a combination of endomorphy and mesomorphy; less than a quarter by ectomorphy.

Looking at different physiques in a population it is apparent that some are more suitable for certain physical tasks than

Table 9
Probable representation of figure types in a population of four hundred people

(a) over 20 of each	(b) No less than 8 no more than 20 of each	(c) No less than 5 no more than 8 of each	(d) No less than 3 no more than 5 of each	(e) 2 or 3 of each	(f) 1 or 2 of each	(g) 1 of each
3.3.4	2.2.6	2.2.5	2.5.2	1.2.6	1.1.7	1.3.6
3.4.4	2.3.5	2.4.5	3.4.5	1.2.7	1.6.3	2.1.7
4.4.3	2.4.4	2.5.4	3.6.2	1.4.5	2.1.6	3.1.6
	2.5.3	3.5.4	4.2.5	1.5.4	2.2.7	3.7.1
	2.6.2	4.2.4	4.5.3	1.6.2	2.7.1	3.7.2
	3.2.5	5.3.2	5.3.4	1.7.1	2.7.2	4.1.5
	3.3.5	5.4.2	5.4.3	1.7.2	3.2.6	4.6.1
	3.4.3			2.3.6	4.5.1	4.6.2
	3.5.2			2.6.1	5.2.2	4.7.1
	3.5.3			2.6.3	5.4.1	5.1.4
	4.3.4			3.6.1	6.2.1	5.1.5
	4.4.2			4.3.5	6.2.3	5.5.1
	4.4.4			5.2.3	6.3.1	5.5.2
	4.3.3			5.2.4	7.1.2	5.6.1
	4.5.2			6.2.2		6.1.2
	5.3.3			6.3.2		6.1.3
						6.4.1
						6.4.2
						7.1.1
						7.2.1
						7.2.2
						7.3.1
						7.3.2
						7.4.1

others. 4—5—2 and 4—5—3 are strong and can carry out tasks requiring good muscle power; and 4—6—2, which is quite rare, has an even more powerful physique. 2—5—3 is a strong physique which can offer more speed. 3—6—2, of which there is only 1%, is quite tall and often has marked athletic abilities, at its best the 3—4—4 is a good all round physique which can offer steady muscle power over long periods.

In Table 9 column (a) includes the more common male figure type, 3—4—4; the bottom of column (b) has the more common female figure type, 5—3—3. Column (b) represents between 25% and 50% of all the figure types in the population.

In recent usage the whole units of the original scheme have been expanded with half units, and 13 degrees are used for each component rather than seven. A somatotype might thus appear as 2½- -4½—4 or 1½—4—4½.

As a system somatotyping needs a good deal of practice; but there is very high level of agreement on ratings between practised users. Detailed knowledge of anatomy is necessary because ratings give scores for skeletal framework, bone widths and lengths, muscle size and fat thickness, as well as comments on the more superficial aspects of body-build.

Differences that do occur between the sexes are accounted for by degree ratings. Sheldon found that although man's ratings are between one and seven, women more rarely reach seven in

Figure 23

3.5.3. This is shown to represent a somatotype common amongst men. This is a strong and muscular physique but not 'chunky'. The muscles show in slight relief. The head is firm with a well modelled face. The neck and shoulders are prominent and well displayed. The trunk exhibits an ample thorax and a neat abdomen and pelvis

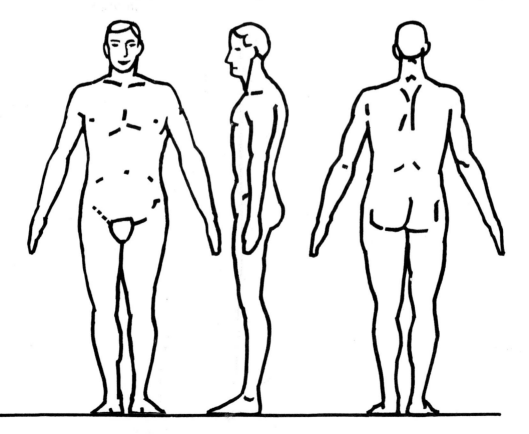

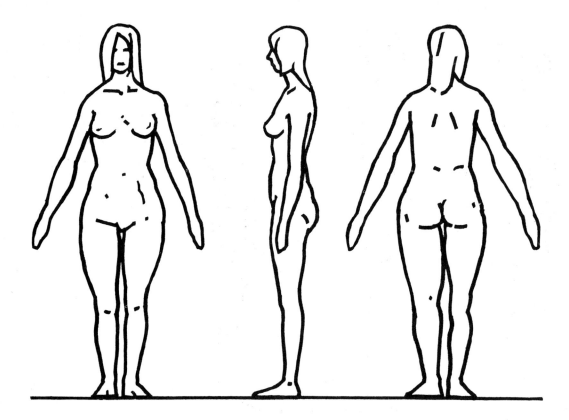

any component. The intervals between ratings remain the same
for both sexes; the ones, twos , threes, fours, fives and sixes of
a component being similar for a man or woman. He found
3—4—4 or 3—5—3 were common ratings for a man, and 5—3—3
common for a woman. In men 5—2—3 or 6—3—3 was rare, and
in women 3—4—4 was rare.

The physique of a person aged between 20 to 30 years offers
the best opportunity for the estimation of an unequivocal rating.
Somatotype rating becomes progressively more difficult to
determine during ageing or when pathological indications are
present. For example, a type of obesity occurring in a predominant ectomorph might produce so much fat as to nearly obscure
the true picture. But examination of the skull, side and front
views of the thorax, the wrists and hands, and the ankles and
feet should reveal the prevailing ectomorph.

Sheldon observed, as did Viola, proportions in some adult
bodily regions that were not compatible with proportions
somatotyped in other regions. They formed disharmonies in
physiques, and were different from the rating which the
majority of a physique might warrant. This appearance of dis-
harmony in a physique he called a dysplasia, a high endomorphic trunk and arms is a typical dysplasia often seen in a
woman. A mesomorphic head with a ectomorphic figure is a
dysplasia sometimes seen in men. See other examples in Figures
25 and 26. Pathological conditions may also present other

Figure 24
5.3.3. This is shown to represent a
somatotype common amongst
women. The head is average to large;
the shoulders small, and the waist
well defined in the centre of the
trunk. Endomorphic softening is
present throughout, though the
three in ectomorphy represents the
longish limbs which are delicate at
the extremities

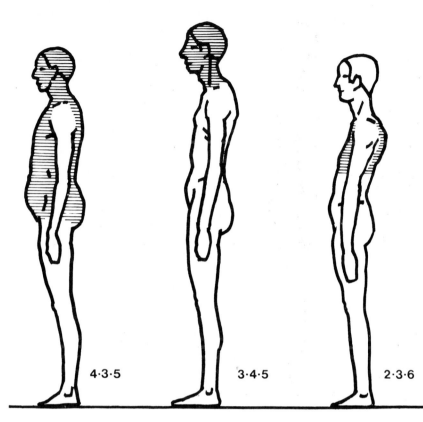

Figure 25
Examples of possible dysplasia in men

4.3.5   An endomorphic head and trunk with ectomorphic limbs

3.4.5   A mesomorphic head and neck supported by an ectomorphic frame beneath

2.3.6   A small physique infected with a collapsed chest

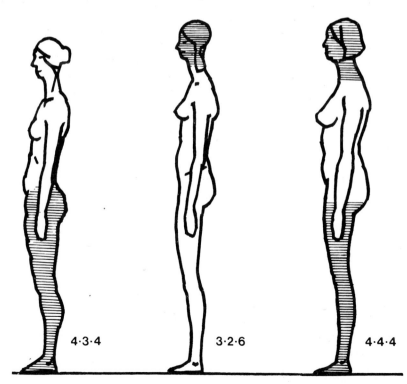

Figure 26
Examples of possible dysplasia in women

4.3.4   Endomorphic legs together with ectomorphic head, trunk and arms

3.2.6   Too small a head for the other parts of the figure which is long and lean

4.4.4   A heavy mesomorphic head, with a mixed endomorphic mesomorphic trunk supported on ectomorphic legs

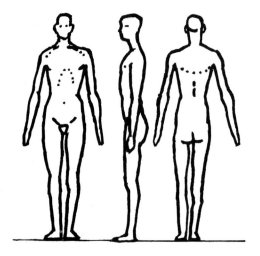

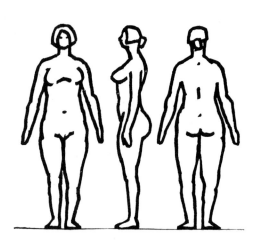

rating 2·3·5

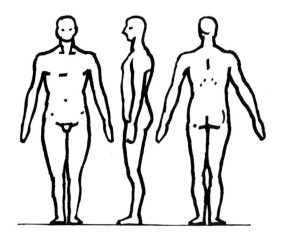

rating 3·6·2

forms of dysplasia. There may also be the presentation of bisexuality in physique. Gynandromorphic scales were created by Sheldon for use in fields of study where a system of measurement was needed for sexing body segments. From the design standpoint it is safe to consider that any bisexuality can be explained in relation to a specific somatotype.

It has been said that it might be possible to evaluate quite a young infant's somatotype when enough data has been collected. The system can be used with some success from a 12 year old onwards. When somatotyping growing children it has been found that the mesomorphic component is often dominant; more particularly so in middle and late adolescence. Body types emerge during the pre-puberal as well as the adolescent

Figure 27

2.3.5   This is a well proportioned physique although the skeleton is light. It is fairly common and although suggestive of power it is not an athletic figure type

3.6.2   Powerful and chunky figure type. Prominent bones and muscles. A physique to be found at athletic meetings

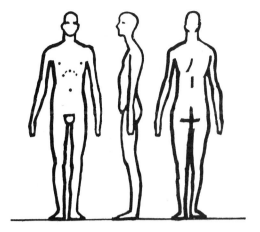
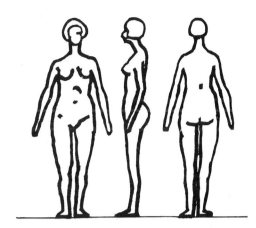

rating 1·4·5

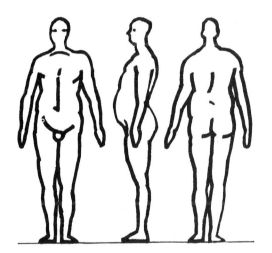
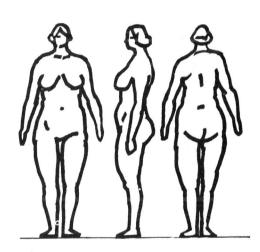

rating 6·3·2

period of growth. There may be slight changes in a developing child's somatotype because of hereditary factors. Therefore, phenotype becomes a more useful term when discussing children's bodily constitution. (Phen derives from the Greek verb *phaino*, meaning to appear. We thus have the description for children; the type that is making is appearance.)

In conclusion it must be said again that somatotype rating distributions will be found to vary from one geographic region to another. Therefore a mass manufacturer of products that have anthropometric constraints must give attention to measurement sampling of the preferred consumer population before a design goes into production: or, alternatively, at the regional stocking selection policy after production.

Figure 28

1.4.5 This shows an interesting combination of mesomorphy and ectomorphy. An unusual and impressive physique

6.3.2 Is one of the giants among figure types. There is a marked endomorphic pneumatic appearance, particularly in the abdomen and the thighs. Although the limbs are well covered with fat in their proximal portions, the limbs are quite heavily muscled. It is a tall type with obvious neck, shoulders and trunk

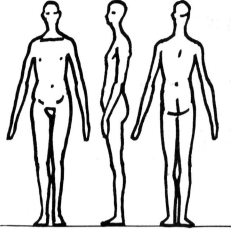

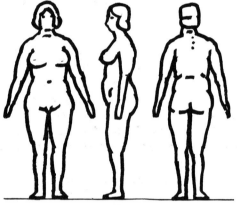

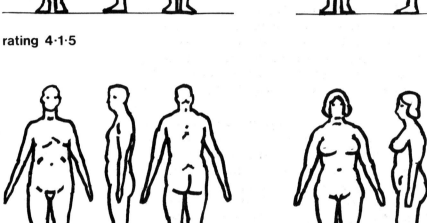

rating 4·1·5

rating 5·4·2

Figure 29

4.1.5   This is a quiet physique with little muscular control. The skeleton is weak and might either be long or short. The female of this type shows the more obvious proportions of her sex, the lack of mesomorphy suiting her

5.4.2   A not uncommon physique. Generally not a tall figure type with fairly large girth circumference measurements

These ratings have been arrived at by observations made during the direction of anatomical studies. They were drawn by the author and it is hoped they caricature the figure types

3 Practical anthropometry

Anthropometry is the practice of measuring the human body. Both the designer and the ergonomist have a continuing need for up-to-date anthropometric data to model equipment, working situations and clothing for optimal use. Static and dynamic anthropometric data will provide the designer with an armature of dimensions around which ideas can grow. Variation of trunk height to leg height or forward arm reach, or alternatively chest girth to hip girth are measurable differences that we can use on behalf of a consumer population.

Static anthropometric data is generally required and used for wholesale clothing manufacture. Dynamic anthropometric data is required for the design of home furniture and fittings in all forms of travel and in a wide range of industrial and engineering, educational and medical activities. It is also required in the design of very specialised clothing which may also be protective or insulating.

There are always a series of body postures that a volunteer subject is instructed to assume for measurement purposes. The body postures are a first attempt at standardization between measurers. The variance discovered in each range of measurements is then directly related to body form and not to its movement. Measurements need to be made from, or between located landmarks on the body surface. Where the landmarks are not bony prominences they must be placed on the skin with a body pencil.

Much information on practical anthropometry is given in the International Biological Programme and this should be consulted on posture, type of measurement, etc, so that universal approaches are known before new research is undertaken.[1]

Data collection
When data is to be collected from a large population, careful organisation is essential to minimise the time taken to measure each subject. An anthropometric sequence must be suitably planned for the speedy taking and recording of the type of measurements required. As well, considerable thought should be given beforehand to the choice of a minimum number of measurements that would give optimum results. The human body is a very complex three dimensional form, and it is extremely easy to indulge in measuring it and go on increasing the measurement score with very little end profit.

Data collecting sheets must be designed in a form suitable for recording in the field, in particular the measurements must be listed in the order that they are to be taken from the subject.

---

[1] See Weiner, J S and Lourie, J A (1969) *Human Biology*, Blackwell

Sufficient information must be recorded to make possible the clear identification of both the subject of the examination, the measurement batch or population group to which they belong, the day of measuring, and the name of the measurer. Data of birth and age in months are also often needed as key information in the evaluation of data.

Some time should be spent in planning the arrangements for reception and measurement, if possible they should be standardised for each station. From filling in the first questionnaire to the final measurement, the progression of individual measuring should be planned for ease and effectiveness. Whatever measuring scheme is adopted it will be found that plenty of space is the first requirement.

Before commencing a survey there are a number of matters that must be given full and careful consideration. Each measurement description must be carefully written up so that it clearly defines the anatomical features and the body marking procedure. It is wise at this stage to discuss these instructions with experienced measurers. The body posture, or position, of volunteer subjects must also have careful consideration. This may involve the director of the survey and an assistant working with a subject for some days prior to the instructions being written up. The author has found that the arrangement of the subject's head in the Frankfurt plane (see Figure 34) can bring significant changes in posture as would be expected, and therefore in measurement result. The quantity of change in measurement result can be quite surprising. This is but one example of the difficulties that can be experienced in composing a posture and then measuring form it.

Finally measurers are given confidence by friendly coaching in the practice period.

The directions for the use of anthropometer, calipers or measurement rig must also be equally well considered. Measurers will depend on them throughout, and particularly during re-training periods.

In summary directions for field work must be as precise and clear as it is possible to get them, but they should not be based only on theoretical knowledge, but on the practical laboratory tests of the director.

In large or small surveys, irrespective of whether measured subjects are arranged in established or new positions, a 'mock up' or trial situation should be created before the actual recordings from volunteers begins and all the subjects' required positions should be photographically recorded.

Measurement procedures should not be undertaken by untrained personnel. Care must be exercised in the selection of measurers, where possible it would be best to use personnel who have some knowledge and experience with the human figure, because accurate measurement requires an experienced knowledge of the superficial tissues that form body shape and

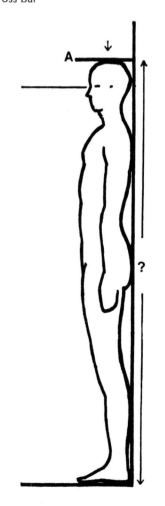

Figure 30
Standing posture for standing height. 'A' represents an adjustable cross bar

individual posture.

Anthropometric methods, even where these are described and rehearsed in great detail, demand a thorough acquaintance with their underlying principles, as well as the technical details and use of instruments. Even simple techniques cannot give reliable results unless the measurer has an understanding of the inherent difficulties and limitations of anthropometric techniques. The formation of a training centre is necessary before any field work is undertaken. Training should commence at least three months prior to the survey. Trainee measurers must work under supervisors until a high degree of fidelity between trainees is achieved. All members of measuring teams should be fully interchangeable, this will insure against any break in measurement due to fatigue or illness once field work commences. Throughout a large survey measurers should have re-training to maintain a necessary high standard in measuring techniques.

Practical measuring

The subject is lightly or partially clothed. Body parts to be marked must be free of clothing. All subjects should be measured at the same time of the day if it is possible. All measures not in the median plane should be taken from the subject's left side unless there is a specific indication for the right or both sides. Weight should be taken on a well serviced beam balance scale machine.

The two most common postures, or measurement positions, are standing and sitting. See Figures 30 and 31.

Figure 31
'**A**' represents an adjustable cross bar

a   represents the basic seated position from which all sitting postures for commercial or industrial use can be derived and measured

b   represents the seated position from which anatomical trunk length can be measured

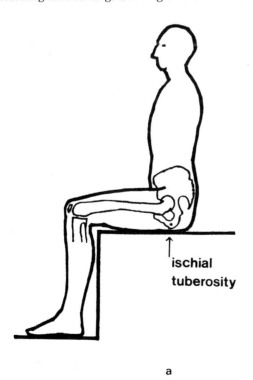

ischial
tuberosity

a

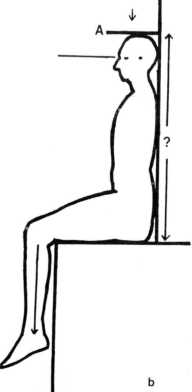

b

Stature or standing height

This measurement must be taken with the subject standing against a vertical background surface in a normal erect position, the shoulders, buttocks and heels lightly touching the background. Account must be taken of the figure type when the body surface is arranged against the background. The arrangement of the head, in the Frankfurt plane, will however take care of the spinal adjustment of the subject to the normal standing posture. The measurement is taken from the vertex of the head to the ground; an adjustable cross bar being brought down to the head and the measure taken from a vertical scale.

Trunk length or sitting height

This measurement is again taken against a vertical scale with an ajustable cross piece being brought down on to the vertex of the subject's head; the measurement being from the head to the horizontal sitting surface. As with the standing posture account must be taken of body physique and the posture controlled by some adjustment of the subject's head, in the Frankfurt plane, and the trunk is as erect as possible.

Trunk length is taken with the legs hanging unsupported. (Figure 31b.) It may be necessary for industrial seating for the feet to be supported. With feet supported trunk posture can be re-adjusted so that it is as upright as possible and sitting height is taken.

If lower leg length is required in relation to sitting height the subject may then be seated on an adjustable seat, and after optimum knee height has been established with the use of the adjustable seat, height from floor to vertex is taken from which seat height is subtracted.

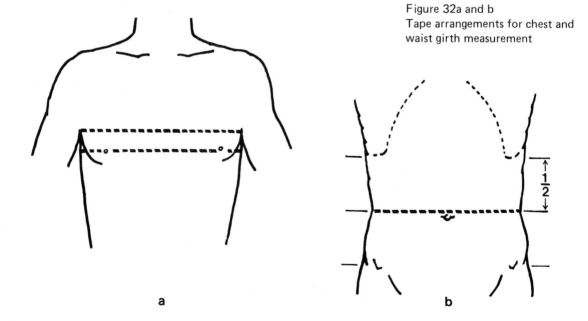

Figure 32a and b
Tape arrangements for chest and waist girth measurement

a

b

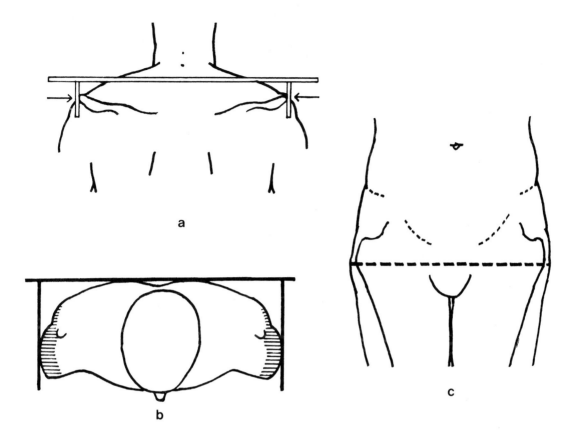

Trunk girths or circumferences

All measurements of the upper part of the trunk should take into account the changes of measurement caused by respiration and this includes measurement of the waist. Three movements in time can be considered; when the chest is in complete expiration or full inspiration, or in a judged half-way stage between the two preceding ones.

Measurements of girths or circumferences can be taken at various levels on the trunk. Some 'fixed points' for levels are armpit height or axillary chest circumference; nipple height or xiphoid cartilage height. (Figure 32.) These measurements can be taken with a linear metal tape measure.

Waist girth or circumference

The subject should be posed in the normal position. The waist measurement will be the least girth measurement recorded between the lower points of the sub-costal angle and the anterior-superior points of the iliac crest. (Figure 32b.) This measurement, as with all other girth measurements, is taken with a light metal tape.

Hip girth circumference

This is taken with a tape at the great trochanters of the femur bones. (Figure 33c.)

Figure 33a, b and c

a Shows lateral points of acromion processes in clear view enclosed by anthropometric rod

b Shows by means of a plan view how most lateral points of acromion process are enclosed by tissue of the deltoid region

c Shows position of great trochanters of femur bones and hip girth circumference measure

Measurements of breadths can be taken on the trunk but they necessitate the use of an anthropometer. This measuring instrument, which is essentially a rod with two arms, one arm of which has a recording counter, requires practice in its use. The practice required is centred largely about the consideration of pressure of the anthropometric arms on the subject.

Trunk measurements include: chest depth; chest breadth; hip width and depth and shoulder widths.

The thighs and legs
Thigh circumference
This measurement can be taken, but is extremely difficult to define in a uniform manner. The subject must stand in a fairly relaxed attitude, although not slumped, with the feet placed evenly on the ground. The tape must be arranged parallel to the ground just touching the lower edge of the buttocks.

Leg length
Leg length is commonly taken as standing height less sitting height. This is easy but only a rough measure. The attempts at leg length measure must depend on the use to which the measure is to be put, for the top of the leg is difficult to determine. Two fixed points for the top of the leg could be the great trochanter, although this is a vague area rather than a point for a linear height measure, or the inferior point of the ischial tuberosity. The crotch offers a third position used in clothing measurement. Leg region measurements include: knee width; calf girth; foot length and breadth; ankle thickness and breadth.

The calf girth is taken as the maximum measure found when the measuring tape is at right angles to the shaft of the leg.

The shoulders
The width of the shoulders is taken from the most lateral point of one acromion process to the other. (Figure 33a.) For industrial purposes this however leaves out of account post-deltoid development with or without subcutaneous fat. The bony width of the shoulders is taken with the adjustable anthropometer. Post-deltoid measurements might have to be taken with the subject seated upright and lightly against an horizontal scale with adjustable arms. (Figure 33b.)

The arms
Arm length can be taken with the arms in the normal position. The length then being from the most lateral point of the acromion process to the tip of the third finger.

Arm span can be measured from third finger to third finger when the arms are fully extended.

Arm girth can be taken halfway between the elbow and the tip of the acromion process. Forearm girth can also be taken.

Arm region measurements include: elbow width; lower arm

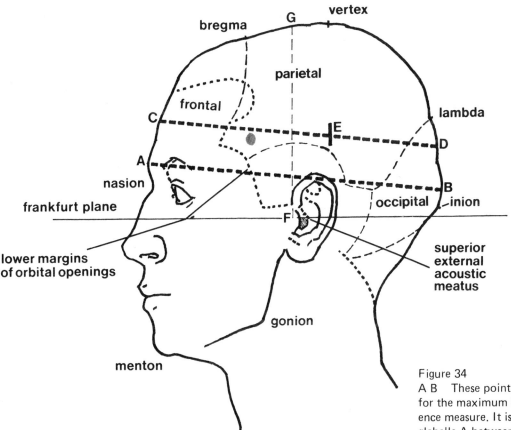

Figure 34

A B These points mark the setting for the maximum head circumference measure. It is taken from the glabella A between the eyebrows round to the furthest projection of the occipital bone B the opisthocranion. Points A B would give maximum head length with spreading calipers

C D Is another circumference measure taken from the frontal eminences C and passing round the parietal bones at their widest point E and round the back of the head from these D

Point E, one each side of the head, the euryon would give maximum head breadth with spreading calipers. F G measure from tragion of one ear vertically over head to tragion of other ear

Glabella, opisthocranion and euryon are three of a number of landmark points on the head used in physical anthropolgy. Various caliper measurements can be taken from euryon, tragion, menton, vertex and opisthocranion

length; hand length and breadth; and wrist thickness and breadth.

The head and neck

Many forms of measurement, devised for the head, would be out of place here. From the designer's point of view their usefulness declines as their number increases. It is likely in the future that helmets or masks for special populations could be designed from stereo-photogrammetric studies of the head and face.

Head circumference can be taken with a metal tape. This can be taken in a line connecting the two frontal eminences and following round the head from these. Other surface measures of the cranium can be devised to solve particular problems of millinery or uniform. Head length and breadth can be taken with spreading calipers.

Neck circumference

This can be taken at the level of the vertebrae prominens when the tape is resting above the head of the sternum at the front of the subject.

A large number of other measurements may be taken. A high proportion concern height from the ground and a number of others include measurements of length or width. Many of them are concerned with measurement dimensions in work spaces and living environments. (See Chapter 5.)

For basic anthropometry the subjects should be without shoes, and holding themselves erect in their natural posture.

Body landmarks

Landmarks are placed on the body to assist accurate measurement. Marks are placed with a skin pencil. Points, small crosses, vertical or horizontal marks are used. The following are some examples:

The head   Calipers, tape or measuring rig would be used. Marking would be by points, eg the tragion.

The neck   Neck base girth. Cross at cervical. Vertical marks at medial clavicle points. Sometimes neck chain used as a guide for more marks.

The trunk   Armscye.   Four points
(1) Midway point between rear acromion edge and front lateral clavicle point.
(2) and (3) Armscye anterior and posterior. Vertical mark ending at intersection of arm with chest and shoulder.
(4) Underarm. Point placed midway across axilla using straight edge.

Midway shoulder point. Placed between lateral point neck base girth and 1st armscye point.

Level of scye depth.   Horizontal mark centre back placed with straight edge between arm intersections.

Waist.   Horizontal waist line mark rear and front. Short marks can be drawn at elastic band level when used.

Abdominal girth.   Side view prominence marked at side with horizontal mark.

The arm   Wrist mark, horizontal tick distal end of ulna midpoint.

Upper arm.   Midpoint between 1st armscye point horizontal mark and elbow point.

Elbow point. Cross, most proximal point of olecrannon.

The leg   Hip level.   Horizontal mark or cross on projection of great trochanter.

Knee.   Cross centre of patella.

Rear Knee. Horizontal mark. Proximal edge of tibia midway between width of femur.

Definitions of measurements from Croney, J (1977), *Man (NS) 12.*

Weight   Taken in light underclothing.

Height   Standing erect with heels together. One measurer checked heels remained on ground and that Frankfurt plane was horizontal. Other measurer held anthropometer vertically and took measurement from floor to anthropometer arm at vertex. Hair style and texture was considered and sufficient pressure was used to brink anthropometer arm to vertex.

Cervical height   Standing erect. Measurement vertically from floor to most prominent spinous process at the seventh cervical vertebra. In many cases subject asked to look down and the process was established by palpation, the measurement taken when the head was erect.

Bust height   Standing erect. Measurement vertically from floor to left nipple prominence and recorded during quiet breathing.

Waist height   Standing erect. Natural waist indent established by fine elasticised cord. Measurement vertically from floor to left centre position on cord below axilla.

Hip height   Standing erect. Measurement vertically from floor to central point of the prominence of the great trochanter of the left femur. Prominence established by palpation.

Tibial height   Standing erect. Measurement vertically from floor to highest point of the glenoid of the tibia.

Sitting height   Sitting erect on table top with feet hanging unsupported, shoulders relaxed. The backs of the knees directly above edge of the table. Head held in the Frankfurt plane, measurement taken between table top and vertex with anthropometer held vertically in contact with subject's back at the sacral and interscapular regions.

Nape to waist   Standing erect. Measurement with the tape from most prominent spinous process at the seventh cervical vertebra in the medial plane to centre back position on elasticised waist cord.

Interacromion width   Standing erect with shoulders relaxed. Measurement horizontally between anthropometer arms held in firm contact with outer edges of acromion process.

Bitrochanteric width   Standing erect with feet together. Measurement between anthropometer arms in contact and level with the greatest lateral trochanteric projections.

Bust girth   Standing erect with shoulders relaxed. Measurement with the tape placed horizontally around chest at nipple level. Measurement during quiet breathing. Care was taken not to constrict the soft tissue.

Waist girth   Standing erect. Tape placed around natural waist indent over elasticised cord. Cord released before measurement was taken.

Abdominal extension girth   Standing erect. Measurement with the tape placed horizontally around body at level of the greatest extension of abdomen.

Hip girth   Standing erect with feet together. Measurement with the tape placed around level given by the greatest lateral trochanteric projections.

Thigh girth   Standing erect. Measurement with the tape passed horizontally around left thigh immediately below gluteal fold.

Calf girth   Standing erect with feet slightly apart. Measurement with the tape passed horizontally around maximum circumference of left calf.

Upper arm girth   Standing erect. Measurement with tape passed round upper left arm at axilla level. Arm away from side whilst tape was placed and returned to side for measurement.

The determination of a sample for an anthropometric survey
The aim should be to state clearly the criteria on which sampling is based so that the identity of the material is established with the least possible ambiguity. A sample has but one use and that is to obtain some insight into the make-up of the population from which it is drawn. Obviously, if this were not the case and each sample were somehow unique, it would be generally useless because of its lack of correspondence to any portion of the population outside of the sample.

One problem in field research is to collect a population sample in the way best fitted to the work in hand, and this will generally require an adequate knowledge of the way a society is organised. If the statistics finally obtained are to be of practical value they must be linked to census data on the distribution by age groups, sex and occupation; the fewer sub-divisions of the population required to accommodate the range of variation which exists in dimensions of the human body, the simpler will be the specifications for the designer.

In most marketing research one can design a sample, by following certain recognised principles, that is fully appreciative of the population requiring investigation. It is not practicable to do this for a detailed anthropometric survey. Because of equipment, space requirements and numbers of personnel it is an economic necessity to accept only volunteers who arrive for measurement. A further important difficulty is that not everyone is prepared to take part in this type of enquiry. A sample, therefore, may not be able to claim to be fully representative.

The following therefore is a summary of factors which would require consideration for the achievement of effective sampling.

> The purposes and use of the measurements.
> How representative does the sample need to be?
> The number and type of the measurements required.
> Regional or socio-economic variants.
> Age, sex, occupation or profession of subjects.

Instrument suppliers

Anthropometer
Holtain Ltd, Brynberian, Crymmych, Pembrokeshire, Wales.
Either a digital read-out Harpenden anthropometer, in carrying case, spare counter, straight and curved branches, lightweight alloy tubes or a wall mounted digital read-out Stadiometer for laboratory use. Both highly recommended.

Siber Hegner and Co Ltd, Talstrasse 14, 8022, Zurich, Switzerland
Siber Hegner and Co Ltd, 8 West 30th Street, New York, NY 10001 USA.
Alternative suppliers of above type instruments in Europe and America.

Caliper (Sliding)
Buck and Ryan Ltd, 101 Tottenham Court Road London W1
Useful addition to above instruments.

Caplipers (spreading or skinfold)
Siber Hegner and Co Ltd, addresses as above.
Good range. Take advice on stock.

Tape, steel or plastic
A Collier (Brixton) Ltd, 423 Coldharbour Lane, Brixton, London.
Chesterman steel tape. Excellent quality.

Weighing machines
Herbert and Sons Ltd, Angel Road Works, Edmonton, London.
Beam scale machine clear for easy 'read-off'. Take advice on stock.

4 The statistical treatment of measurements

When man and his material needs are the objects of study it is more convenient to consider him *en masse* in terms of populations.

Populations may be surveyed and sampled as a whole or in specific sections of users. Populations can be defined fairly readily in relation to a design project under consideration. In fact the design project itself more often than not chooses the population. Dentists' chairs, school desks and cooking ovens pick their own victims. Using simple standard statistical methods it is possible to establish useful facts about a chosen section of a population. Calculation of an average weight; or an indication of the range of weight by giving the lightest person and the heaviest person are two answers easily discovered. The examination of a series of weights to divide it into classes of lightweights, medium weights or heavy weights is a third obtainable answer. Each answer in statistics depends on the manner in which the question is framed. In applied anthropometry questions about human measurements are prompted by design projects. The ultimate usefulness of any anthropometric survey depends completely on the extent to which the total body of derived data can be transformed by statistical analyses into summaries or 'key' dimensions which can be used for solving design problems. The fact that the questions must be asked in terms of quantitative measurements is an initial assurance of success. Applied anthropometry excludes qualitative biological differences, eg different hair colours or physiological changes with age. It also excludes simple numerical differences in anatomy, eg numbers of muscles or numbers of digits. Quantitative measurement does include alterations of stature, body structure and muscle power during a life-time. All measurable human quantities can be graded between two extremes, although the majority of measurements cluster somewhere in between the two extreme values, and with enough statistical evidence total range ratios can be calculated. As a measure of dispersion the range ratio does have the drawback of being completely dependent on the number of recordings made, and by simply increasing them the range could increase; it is therefore unstable from sample to sample. It does retain a utility in the description of small samples.

Table 10

Some examples of total range ratios

Statue (adult)	1.25 : 1
Total arm length (adult)	1.5 : 1
Total body weight (adult)	2.5 : 1

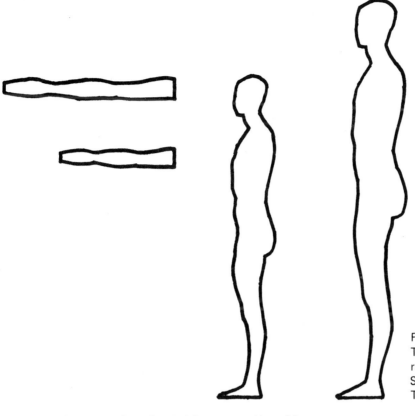

Figure 35
Two examples of human range
ratios in diagrammatic form
Standing height 1.25 : 1
Total arm length 1.5 : 1

It can be seen that the total range ratio of human measurements is comparatively narrow against the possible wide range of dimensions the designer can achieve when he creates a new design. (Figure 35.) The lesson for the designer here is that, detailing, and the careful consideration of slight adjustments to measurements, when designing a product for human use, are all important.

Careful thought must be given to the type of measurements taken from a population. Measurements must be relevant to the problem to be solved, eg in the case of a seated operator: is it forearm- and hand-length or depth of trunk and thigh length that is wanted or both? (Figure 36.)

It is therefore necessary to know the measurements or measurement differences required for the use of a particular piece of equipment or garment. If it is found that a measurement can be invalidated from any point of view it is better to exclude it. When the type of measurement required has been decided it should be taken several times on each chosen subject. It is preferable for more than one person to take all the measurements. Although such comparisons are time consuming, they are well worth while. Errors must be reduced to a minimum so that each measurement is totally reliable.

When we define the scope and limits of a population for a particular purpose, sooner or later we find ourselves confronted with the average person. The average person is a concept of a typical person that emerges from the consideration of a number of persons, but it is nevertheless no one in particular. The average person is the general representative of a quantity of persons. It may be said that most men are taller than women, but of course some women are taller than some men. However, we can say the average man is taller than the average woman. In some fields of work simple everyday facts about a population that can be represented by an average person may be enough. A man rather than a woman; a factory worker rather than an office worker. To design for the average person, however, can lead to dangerous errors. Very few persons in a population are average in a large number of measurements of bodily dimensions or capacities that could be examined in an anthropometric study. Percentages of a population may be inadequate in some dimensions or capacities and not in others. If we pursue the average in terms of more and more dimensional measurements we find that as the total number of dimensional measurements taken increases so the percentage of the 'average' person who can represent them also decreases.

The average person nevertheless remains a useful concept in statistical application. Anthropometry is the empirical study of man by measurement. Statistical methods enable us to examine ranges of measurements and make observations about their frequency of occurrence and distribution: statistical procedures also reveal significant trends and relationships in collected data that have direct application to design problems.

All the dimensional measurements we can take on a person vary from most other similar dimensions measured on a number of other people. In fact we find that the variety of one measurement increases as the sample total increases. Eventually we discover duplicates and so on, for the increase is not indefinite.

Two facts emerge from this statement: the range of human measurement is surely a matter of chance; and quite a large number of measurements have to be taken before we can have an understanding of their variation.

When a large sample of one kind of measurement of body dimension, say height, has been selected from a population a pattern can be observed in their distribution if they are listed in order of scale and number. If a graph line y is then plotted of the scale of the measurements against their frequency of occurrence we have as a result a symmetrical or bell-shaped looking curved line.

A curve produced in this manner from a quantity of varying measurements is called a Normal curve or curve of Normal frequency distribution. (Figure 37.) This curve is one often met with in experimental biology where many continuously variable organic forms are being studied.

Because a distribution of this type lends itself so happily to mathematical treatment it forms the basis for much work in

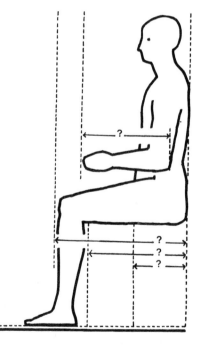

Figure 36
Relevant measuring is important in two ways. It necessitates the formulation of a planned approach and the careful consideration of detailing. It also eliminates unnecessary measuring which wastes time and money

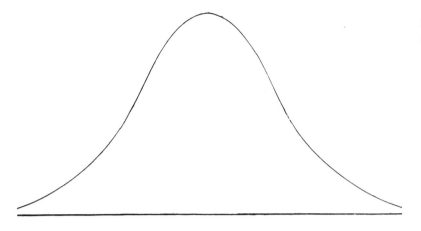

Figure 37
A Normal curve

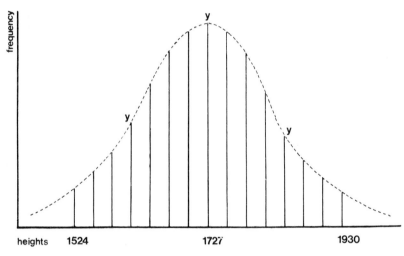

Figure 38
Men's height in millimetres, their frequency of number plotted against their range of measurement from the shortest to the tallest giving a graph line **y**. Human heights are a good example of a dimensional measurement which when taken at random comes very close to a normal distribution

theoretical and applied statistics. Where factors are a matter of chance, as is the case with man's dimensional measurements, some measurements will appear more frequently than others. The more frequent scores will tend to cluster round a centre or mean value. Other measurements of more extreme value and therefore rarer will be spread out on either side of the centre or mean value. (Figure 39.)

The centre or mean value, or the value which most frequently occurs, could be the value of measurement which represents the average man. There are however three different types of average which may or may not coincide; they are the mode, the median or the arithmetic mean. The mode is the value which most frequently occurs in a chance collection of measurements. So the mode is the representative of the commonly used average. To say that an average man is 1727 mm (5 ft 8 in.) in height is to say that more often than not a man is about that height. But it does not say that the mean of all men's height is 1727 mm. The median is the value that is the centre item in a chance collection of measurements. To say that the median of a range of men's heights is 1727 mm, is saying that 1727 mm is the value of the height half way through a number of values of men's heights arranged in order of scale.

The arithmetic mean is the measurement value found after all the values have been added together then divided by the total number of value occurrences. To say that the arithmetic mean of men's heights is 1727 mm is to quote the value given from a number of men's heights when added together and divided by the number of men measured.

In a perfect Normal curve distribution the three different

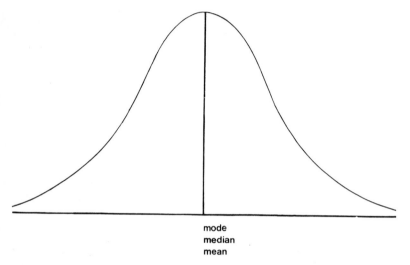

mode
median
mean

Figure 39
Showing the mode, median and the arithmetic mean at the centre of a distribution. If a distribution is perfectly normal and produces a symmetrical curve all three forms of the average will coincide at the centre of the curve

types of the average coincide with one another. It must be admitted that the exact symmetrical curve shown is an idealized version, but many body measurement variables, as well as height, come so near to a normal distribution that they may be treated as such.

It is the interplay between genetic factors and environmental factors that determines the continuous variation of human measurements: determines, in fact, that their distributions are a matter of chance.

When we need to take a sample of measurements produced by chance, care and thought must be given to the manner in which we select the sample. To obtain a Normal curve that will yield useful results a sample of measurements must be both reasonably large and unselective. For a sample to be unselective it is necessary to devise procedures that are without bias, and that offer every subject of a population that is to be surveyed an equal chance of being chosen for measurement. It is not within our scope to discuss advanced sampling procedures here, but they are recommended for further study. Thorough sampling necessitates collecting a sufficient quantity of data that will show the full ranges of particular measurements in a population. It is not possible or practical to use a whole population of users, especially if it is a large population. There is however a minimum number for a sample to be of any use. A hundred

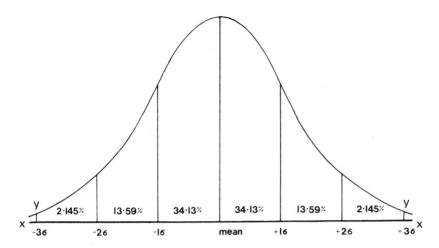

Figure 40
The area under a normal distribution divided by standard deviations

Here **y** and **x** enclose an area representing a whole sample. Perpendiculars enclose portions of the area

σ = standard deviation

The areas enclosed by each deviation are shown

68.26% of values in any normal distribution come within ± one standard deviation

persons is the very least number likely to produce a result that would be worthwhile. Three to five hundred is better, larger samples can be taken, but the collaboration of a team of investigators would then be required which requires organisation of a high order. If very small samples only are obtainable assistance must be sought for statistical methods to establish their reliability.

In general design problems concerned with body extremities require smaller samples than do the large body dimensions. The range of measurement of standing height or weight is much greater than the range of measurement for smaller anatomical details and consequently requires greater size samples.

The Normal curve is important, because it is so convenient for the assembly and processing of new data, and therefore generally forms the basis for all statements of probability. This remains true despite a rather chequered story in mathematical history which dates back to 1733 when it first saw the light of day in print, as a result of the indefatigable study of Abraham de Moivre. If we take the area enclosed by the Normal curve as 1 or 100% we can see that smaller areas of it can predict probabilities in any binary event. (Figure 40.) De Moivre calculated a fraction of the total area under the Normal curve each side of the mean line to a distance of ±1 which gave a value of 0.682688. ±1 being one standard deviation each side of the centre line of the score.

If vertical lines are erected at ±1 from the mean an area of 68.26% is enclosed or 34.13% for each deviation. Calculations for two standard deviations give values under **y** of 0.2718 or 27.18% more area enclosed; 13.59% for each deviation.

Calculations for three standard deviations give values of 0.04290 or 4.29% more are enclosed; 2.145% for each deviation.

The device of the standard deviation is a most useful statistic when considering the distribution of chance occurrences. When

[1] See Weiner, J S and Lourie J A (1969), *Human Biology*, Blackwell

there is a Normal distribution we can say there is about a 33.3% chance of having a measurement outside one standard deviation, and about a 5% chance of having a measurement outside two standard deviations.

Standard deviations allow statements to be made about samples of measurement taken at random from a population. For example, if 120 measurements are taken of men's heights we know that about 80 of the men's heights lie within ± one standard deviation, and that about 114 lie within ± two standard deviations. By using the standard deviation equation we may determine the dispersion or variability of a sample about its mean value. The value of a variable in a Normal distribution would therefore be the arithmetic mean of all the variables plus or minus the value of the number of standard deviations. It is usual to present anthropmetric results in terms of the arithmetic mean and one value of standard deviation, this presents succinctly a reasonable idea of the variability of the sample. If this is presented with similar data of another time or place it can also express a historical or bio-metric relationship (see Table 11). When most measurements cluster near the mean value the standard deviation will be small; on the other hand if many of the measurements are much smaller or much larger than the mean value the standard deviation value will be larger. The equation for the standard deviation calculation is:

$$SD = \sqrt{\sum (x - \bar{x})^2 / N}$$

$\sum$ being a mathematical symbol meaning that all the values in brackets should be added together, x represents all the individual measurements in the distribution, $\bar{x}$ represents the arithmetic mean of all the values in x : $\bar{x} = \sum x/N$, N representing the total of individual measurements in the distribution. Variability of body measurements is related to the type of dimension taken. Thus direct body measurements such as girth circumference have a greater variability about the mean than indirect body measurements such as total height or leg length.

Each type of dimension is subject to some error, this can be calculated and is called the standard error. The standard error relates to the statistic in question, the size of the sample and in most cases to the standard deviation of particular data. The standard error calculation may be met because it has extensive use in industrial and commercial fields, however for reasonably large anthropometric surveys standard errors either of the standard deviation or the mean are very small and have little significance.

Often designing a product for a consumer population uses means selecting to work in relation to values of measurement that come within the areas of one or more standard deviations. Two standard deviations contain the spread of the central 95%

Table 11
Comparisons between means and standard deviations

	Means		Standard deviations			
	Croney	*Kemsley*	*Croney*	*Kemsley*	*Forziati*	*O'Brien*
Weight	54·7 kg	57·5 kg	6·78 kg	8·34 kg	7·63 kg	8·93 kg
Height	163·0 cm	161·1 cm	6·62 cm	6·29 cm	6·68 cm	6·29 cm
Cervical height	138·9 cm	138·0 cm	6·04 cm	5·81 cm	6·15 cm	5·89 cm
Hip height	82·3 cm	80·9 cm	4·89 cm	4·23 cm	6·37 cm	4·55 cm
Tibial height	42·2 cm	43·4 cm	2·51 cm	2·56 cm		
Nape to waist	39·0 cm	38·2 cm	2·48 cm	2·23 cm		
Interacromion width	36·5 cm	35·0 cm	1·76 cm	1·82 cm		
Bitrochanteric width	33·5 cm	33·0 cm	1·89 cm	2·03 cm		
Bust girth	86·2 cm	89·5 cm	4·90 cm	6·10 cm	5·79 cm	6·90 cm
Waist girth	63·9 cm	64·0 cm	4·36 cm	5·61 cm	5·35 cm	7·13 cm
Abdominal ext. girth	80·0 cm	85·0 cm	5·72 cm	6·63 cm	7·08 cm	8·60 cm
Hip girth	93·6 cm	95·2 cm	5·17 cm	6·08 cm	5·72 cm	7·01 cm
Calf girth	33·9 cm	34·3 cm	2·66 cm	2·39 cm	2·54 cm	2·74 cm
Upper arm girth	26·1 cm	27·4 cm	2·89 cm	2·64 cm		

of values from the mode. This would be excellent coverage and in practice something rather less than this is achieved.

It is ofen useful to divide a distribution of measurements into groups or classes. The number of classes must finally be decided from the quantity and variation of the measurements in a distribution. A number below 10 might be used; but about 10 to 20 show good results.

When choosing the number of classes the number of measurements that each class contains is also being decided. If a distribution contains 400 measurements and it is divided into 20 classes then each class will contain 20 measurements. The number of classes chosen must finally be a matter of personal experience and judgement.

The following is an example of how this might be done. Supposing a survey of men's standing heights had produced a distribution of measurements ranging from 1473 mm (58 in.) to 2000 mm (78 in.). The first step towards determining the number of class groupings is to find out the maximum quantity of measurements that could occur between these minimum and maximum values. Altogether there is the possibility of 528 values. Class groupings can be evaluated with the knowledge that between 10 and 20 classes are the minimum and maximum number of classes that show a useful working result. A range of 528 mm divided by 20 would give a change of height through each class of 26.4 mm, a little more than one inch. This for design purposes would probably be too fine. A division by 11 produces a good result, so that all the classes are balanced within themselves, and 11 classes also brings about a useful increase of significance. The 11 classes would appear as follows:

mm	in.	
1473—1520	58.0- -59.8	
1521—1568	59.9—61.7	
1569—1616	61.8--63.6	
1617—1664	63.7—65.5	
1665 -1712	65.6—67.4	
1713--1760	67.5--69.3	←This class contains the mean
1761—1808	69.4—71.1	
1809—1856	71.2—73.0	
1857—1904	73.1—74.9	
1905—1952	75.0—76.8	
1953—2000	76.9—78.7	

The division of the data into 11 classes would mean that each class contained 48 values of the increase in height. The difference in standing height through each class would be 48 mm (1.9 in.).

The final step is to tally the actual measurements with their correct class grouping. Each measurement must be placed within its class limits. The 1473 mm (58 in.) measurement is scored into the first class with a tick; the 2000 mm (78.7 in.) measurement is scored into the last class. A 1920 mm (75.6 in.) measurement would be placed in the 1907.5 mm (75 in.) to 1952.8 mm (76.5 in.) class, and so on until all the measurements are tallied in.

```
1473—1520   I
1521—1568
1569—1616
1617—1664
1665—1712
1713—1760
1761—1808
1809—1856
1857—1904   I
1905—1952   I
1953—2000
```

The final scoring might look like this:

							Fc
1473—1520	┼┼┼						5
1521—1568	┼┼┼	┼┼┼					10
1569—1616	┼┼┼	┼┼┼	┼┼┼				15
1617--1664	┼┼┼	┼┼┼	┼┼┼	┼┼┼			20
1665--1712	┼┼┼	┼┼┼	┼┼┼	┼┼┼	┼┼┼		25
1713—1760	┼┼┼	┼┼┼	┼┼┼	┼┼┼	┼┼┼	┼┼┼	30
1761--1808	┼┼┼	┼┼┼	┼┼┼	┼┼┼	┼┼┼		25
1809—1856	┼┼┼	┼┼┼	┼┼┼	┼┼┼			20
1857—1904	┼┼┼	┼┼┼	┼┼┼				15
1905—1952	┼┼┼	┼┼┼					10
1953—2000	┼┼┼						5

Note fifth tallies are marked diagonally across the preceeding four. The tallies are then placed as a figure in a column on the right which is the frequency column.

The value of grouping according to frequency in classes is that it makes it possible for the mind to conceptualise a random collection of figures and to place interpretations upon them.

It will be seen from the accompanying example that grouping measurement data into classes establishes ideas about data that could be gained in no other way, but at the same time the individuality of the measurements has been lost. It is in the consideration of the loss and gain of information that experience is required when choosing the number of classes for a distribution.

The construction of histograms is helpful as a visual aid to display chosen class divisions. Histograms show a series of rectangles for each class in a frequency distribution. A histogram is a graph which uses rectangles as building bricks forming steps. (Figure 41.) The width of each rectangle gives the limits of the measurement's frequency of occurrence. The area of each rectangle is representative of the number of measurements in each class.

Histograms can also serve as a useful means of determining the importance and significance of class groupings.

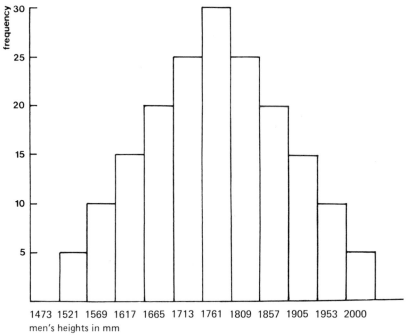

Figure 41
A histogram constructed from the invented data of men's standing heights in the text

Another useful aid for studying the distribution of body measurements for design purposes is a graph cumulative frequency distribution.

Add up the frequencies class by class from a frequency distribution. Using the material from the previous example we have:

Frequency	Cumulative Frequency
5	5
10	15
15	30
20	50
25	75
30	105
25	130
20	150
15	165
10	175
5	180

Which would produce a total of 180 frequencies.

To construct a graph with the cumulative frequencies, the frequencies would be placed on a perpendicular axis from one to 180 and the classes of the distribution on the horizontal axis. The cumulative frequencies can then be plotted against the appropriate class limit. (Figure 42.)

Cumulative frequency distribution graphs will give cumulative percentages per population and offer percentile values as well.

Percentiles

Percentile values are found by dividing any number of quantities

Figure 42
A cumulative frequency distribution graph. The right hand portion of the graph has been set off one extra division which is marked off in cumulative percentages or percentile values.
To establish the percentile value of a height: Decide where the height is situated in a particular class. Produce a perpendicular line from this point to the curve above, the read off from the right the percentile value which is horizontally opposite this curve intersection. Cumulative frequency curves are also called ogives. Cumulative distributions are useful and easy to consult for design purposes because they show percentages per population so clearly

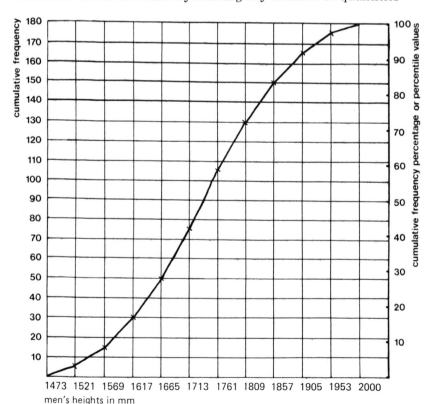

men's heights in mm

into 100 equal groups in order of their size, from the least to the greatest in an orderly sequence. So extreme values of size, very small or very large, are represented by small or large percentile figures; the smaller or larger the percentile figure the slighter the chance of the occurrence of that value.

Percentile values are given in fractions of 100 in relation to an array of measurements. If the measurements in an array are ungrouped, finding the percentile merely necessitates counting up the number of measurements.

0 and 100th percentiles are the smallest and largest measurements in a distribution. If 1473 mm (58 in.) is the smallest height and 2000 mm (78.7 in.) is the largest height, the 0 percentile is 1473 mm (58 in.) and the 100th at 2000 mm (78.7 in.). The 10th percentile is therefore found by counting off from the lowest measurement one tenth of the measurements; the 20th percentile by counting off one twentieth of the measurements, and so on. If an array of measurements is grouped by frequency it will be necessary to use a process of interpolation.

Percentile values show the cumulative frequency of occurrence of measurements in order of their size as a percentage of the population being studied.

To find a percentile value when data is grouped into classes in millimetres proceed as follows

1473 –1520	5	5
1521—1568	10	15
1569—1616	15	30
1617—1664	20	50
1665—1712	25	75
1713—1760	30	105
1761—1808	25	130
1809—1856	20	150
1857—1904	15	165
1905—1952	10	175
1953—2000	5	180

To find the 10th percentile:

First find a tenth of 180 the total frequencies $\frac{1}{10} \times 180 = 18$

Count up the frequencies from the smallest end of the distribution until the 18th is reached.

The 18th comes within the third class with 15 frequencies.

We require three of the 15 frequencies to make 18.

To find the value of 1 frequency in this class divide 48 mm (1.9 in.), the size of the class, by 15.

This equals 3.2 mm (0.126 in.).

So 3.2 mm × 3 added to 1569 mm (the smallest end of the third class) will give us the 10 percentile value.

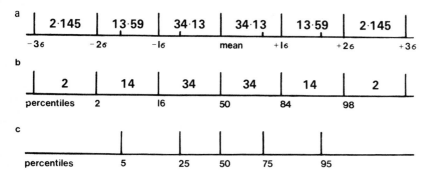

a

2·145	13·59	34·13	34·13	13·59	2·145

−3σ   −2σ   −1σ   mean   +1σ   +2σ   +3σ

b

2	14	34	34	14	2

percentiles 2   16   50   84   98

c

percentiles   5   25   50   75   95

3.2 mm × 3 = 9.6 mm

1569 mm + 9.6 mm = 1578.6 mm

Say 1579 mm = 10th percentile.

From an array of height measurements we could find the value of height that would include 90% of the population. This information for example could be used to give the height for a door, that would only inconvenience the remaining 10%. A value for a 'reasonable fit' could be obtained for other products in a similar manner. The dimensions for the design of products wherever possible should be derived initially from selected percentile values based on the distributions of users' measurements.

The mean and standard deviations of a Normal distribution can be given in percentile values. (Figure 43a, b, c.) Beyond one standard deviation percentile values can be seen to act as cut-off values to exclude the very smallest and largest size measurements which can rarely be fitted into a design scheme. One standard deviation each side of the mean value gives percentile values of from 16 to 84. (Figure 43b.) So that the 5th and 95th percentiles become cut-off values within ±2 standard deviations. Some percentiles have their own names, the 50th percentile is the median; the 25th, 50th and 75th are quartiles, and there are nine deciles the 10th, 20th, 30th, 40th and so on.

Correlation and regression analysis

Many anthropmetric problems that arise in design or design management cannot be solved by the determination of percentile values, although clearly they have great value when only one measurement variable has to be considered. More often what is required is the analysis of the relationship of two or more variables. The following is an outline of the statistical procedures that may have to be used to do this, procedures which are generally brought together in texts on multi-variate methods. It has not been thought appropriate to deal with mathematical analysis in depth here; for example the proper consideration of linear regression requires ample space. However, the books in the statistics section of the reading list are highly recommended for those who can extend themselves mathematically.

Figure 43a, b, c
This diagram shows a comparison between a Normal distribution divided by standard deviations and a linear comparison of percentile values

a   represents the base of a Normal distribution and the intervals of standard deviations

b   represents a percentile scale marked with percentile values that relate very nearly to the ± deviations on a, the second percentile relating to −2σ, the sixteenth percentile relating to −1σ and so on. The percentage of values represented under the Normal curve between −1σ and −2σ 13.9% is very close to the fourteen percentile difference between the second and the sixteenth percentiles. Other percentile values can be compared in a similar manner. Comparisons between a and b is another way of demonstrating that percentile values give percentages of a distribution of a population. Percentile values can replace standard deviations, but are not so accurate.

c   represents a percentile scale marked with well used percentile values and their relationship to ±2 standard deviations is marked off on a

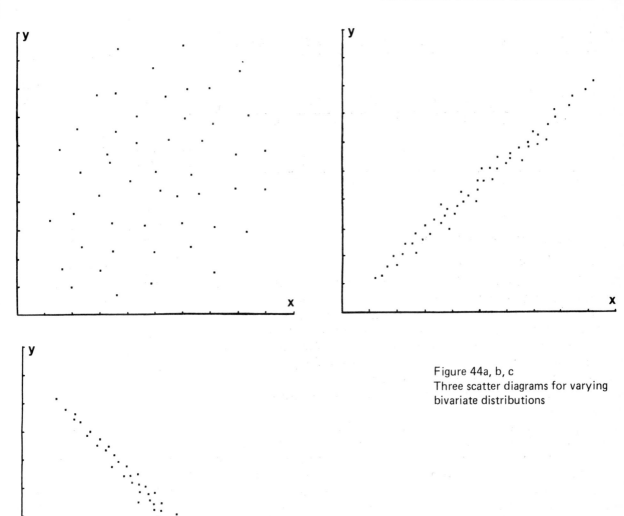

Figure 44a, b, c
Three scatter diagrams for varying
bivariate distributions

To the designer in general it must be said that it is sufficient
if a reasonable understanding of multi-variate methods and
applications is obtained up to the level required for controlling
clear graphic displays of statistical analysis for office and work-
shop use. The designer or design manager is now often one of a
team working towards a desired goal. The team work will often
generate a large amount of anthropometric data; because of this
fact and the efficiency of hand calculators and modern comput-
ing machines in dealing with the analysis of many variables, the
designer will usually find that he has the assistance of a statis-
tician or computer programmer. It is, however, left to the

designer to determine the significance of the computed findings.

Scatter diagrams are a good first method for showing the intensity of the relationship between any two distributions, because they give a simple visual idea of correlation value. So they are quite often used at an early planning stage. Depending on the nature of the distributions it will be seen that there is either a more or less positive association between them; if there is a more positive association the scatter will follow a defined direction. Figure 44 **a** shows a less positive association, **b** a more positive association; and in this case any point in the scatter indicates that for any increase in value of **x** on the abscissa there will be some increase in the value of **y** on the ordinate. It should be noted that as shown in **c**, it is possible to have a negative association.

Figure 45 is an example of a scatter diagram being built up from men's crotch heights. The intensity of the association has a definite direction, and it is suggested that it may come near to being linear.

Although it is now generally well known which body measurement variables have a more positive association in European samples there are still advantages in using scatter diagrams. If it has been determined how many subjects in a survey come at each scatter point numerically, then there is a method to hand

Figure 45
By plotting men's crotch heights against their standing heights the intensity of their association could be studied. If both the distributions were Normal the greatest intensity of correlation would be about the area where the more central values of the distributions met. The crosses show how the intensity of correlation might build up. The arrow shows the expected diagonal direction they would make. The more distinct the diagonal pattern, the more close the relationships of the two variables

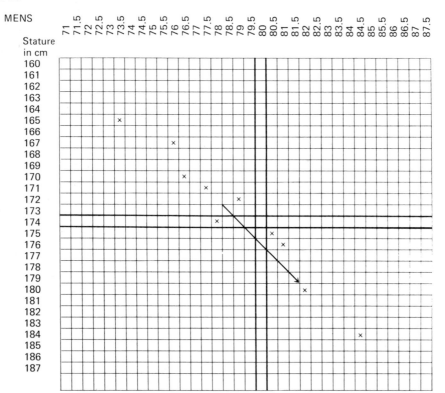

for establishing a percentage of possible users within the constraints of the two distributions. To do this scatter diagrams are converted to bivariate distributions. (See Figure 46 and discussion.)

The co-efficient of correlation gives the measure of the intensity of the relationship between two variables. Any co-efficient value above zero indicates some degree of functional relationship, if the co-efficient value were to reach 1.0 or unity a perfect functional relationship would exist. Meaning that as one variable increases the other increases a 100% of the time. Correlation co-efficient (written as r) = 1. In practice the relationship of anthropometric variables always falls short of unity. For example the relationship between total height and cervical height could be 0.978 (see Table 12) and for this example means that 95% or more of the heights of these young women could be predicted from their cervical height or vice-versa.

The formula for the co-efficient of correlation is

$$r = \frac{1}{N} \sum \frac{(x - \bar{x})}{\sigma y} \frac{(y - \bar{y})}{\sigma y}.$$

It is possible to have two values of r that are the same,

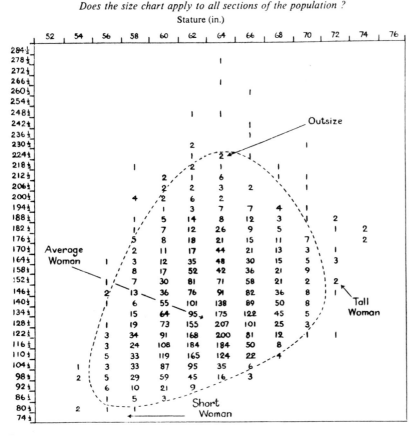

Does the size chart apply to all sections of the population ?

Figure 46
This shows a bivariate distribution of stature and weight. The variables from the distributions associate in pairs; each pair of variables re-occurs the number of times subjects are found that have two particular body dimensions in common. Each scatter distribution in this form would represent a particular co-efficient value. The use of such a distribution is that it can indicate percentages of the sampled population. Other contours could be drawn other than the 99% contour shown.

This particular example shows how a sample of about 5000 women are distributed over the weight and stature field. From it information can be extracted on the nature of four possible size groupings for clothing purposes; short women's, tall women's, average women's and outsize women's. The analysis of other measurement dimensions (dependent on weight and height) taken on the women, would allow conclusions to be drawn for the construction of size charts
From Kemsley, W F F (1957), *Women's Measurement and Sizes*, HMSO

Table 12a
Data on 311 young women fashion college students; means, ranges and standard deviations and 5th and 95th percentiles from Croney, J (1977) *Man* (NS) 12

	Mean	Range	S.D.	5th	95th
1 Age	235·57 mths	206–315 mths	21·42	224·00	262·00
2 Weight	54·7 kg	39–85 kg	6·78 kg	45·2 kg	67·1 kg
3 Height	163·0 cm	141·5–183·0 cm	6·62 cm	152·3 cm	173·6 cm
4 Cervical height	138·9 cm	119·5–156·6 cm	6·04 cm	129·8 cm	149·9 cm
5 Bust height	118·7 cm	100·0–137·1 cm	5·82 cm	109·8 cm	128·0 cm
6 Waist height	101·4 cm	85·0–120·4 cm	5·34 cm	94·3 cm	110·0 cm
7 Hip height	82·3 cm	67·5–98·0 cm	4·89 cm	75·8 cm	90·6 cm
8 Tibial height	42·2 cm	35·1–50·2 cm	2·51 cm	38·5 cm	46·3 cm
9 Sitting height	85·9 cm	75·3–97·5 cm	3·60 cm	79·6 cm	90·7 cm
10 Nape to waist	39·0 cm	32·6–45·9 cm	2·48 cm	35·9 cm	42·7 cm
11 Interacromion width	36·5 cm	31·1–42·2 cm	1·76 cm	34·4 cm	39·2 cm
12 Bitrochanteric width	33·5 cm	28·8–41·6 cm	1·89 cm	31·1 cm	36·5 cm
13 Bust girth	86·2 cm	76·3–110·9 cm	4·90 cm	80·1 cm	94·2 cm
14 Waist girth	63·9 cm	47·0–85·5 cm	4·36 cm	59·5 cm	70·6 cm
15 Abdominal ext. girth	80·0 cm	63·4–101·0 cm	5·72 cm	72·0 cm	98·4 cm
16 Hip girth	93·6 cm	82·3–115·5 cm	5·17 cm	86·1 cm	102·4 cm
17 Thigh girth	52·8 cm	42·6–66·3 cm	3·79 cm	47·7 cm	58·8 cm
18 Calf girth	33·9 cm	28·0–43·6 cm	2·66 cm	30·4 cm	39·0 cm
19 Upper arm girth	26·1 cm	20·3–34·7 cm	2·89 cm	22·9 cm	29·5 cm

Table 12b

Measurement	1	2	3	4	5	6	7	8	9	10	11	12	13	14	15	16	17	18
2	·110																	
3	·096	·558																
4	·088	·543	·978															
5	·101	·485	·950	·949														
6	·081	·512	·884	·869	·878													
7	·097	·412	·824	·825	·824	·874												
8	·119	·432	·822	·829	·835	·839	·822											
9	·078	·492	·779	·745	·725	·635	·549	·518										
10	·025	·376	·546	·572	·496	·340	·303	·349	·612									
11	·056	·530	·435	·417	·376	·402	·381	·418	·310	·192								
12	·109	·710	·434	·430	·375	·391	·266	·337	·478	·300	·483							
13	·150	·725	·252	·244	·199	·228	·196	·185	·242	·184	·448	·494						
14	·161	·737	·297	·290	·258	·283	·232	·228	·311	·229	·411	·589	·770					
15	·105	·743	·342	·347	·295	·288	·224	·233	·386	·318	·355	·661	·716	·754				
16	·082	·768	·335	·319	·281	·279	·167	·210	·427	·313	·385	·798	·630	·668	·755			
17	·109	·761	·250	·239	·211	·239	·091	·167	·333	·221	·390	·731	·627	·626	·667	·823		
18	·048	·762	·417	·409	·330	·372	·297	·293	·470	·320	·349	·646	·527	·567	·543	·768	·712	
19	·082	·723	·185	·179	·125	·203	·141	·136	·233	·128	·465	·634	·726	·720	·709	·759	·753	·645

Measurements, and other results from them, named in Table 12a on facing page

J. E. CRONEY

MEASUREMENT GROUPINGS DEFINED BY THE DEGREE OF CORRELATION

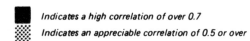

		2	13	14	15	16	17	18	19	12	11	3	4	5	6	7	8	9	10
Weight	2																		
Bust girth	13																		
Waist girth	14																		
Abd.ext. girth	15																		
Hip girth	16																		
Thigh girth	17																		
Calf girth	18																		
Up.arm girth	19																		
Bitrochanteric width	12																		
Interacromion width	11																		
Height	3																		
Cervical ht.	4																		
Bust height	5																		
Waist height	6																		
Hip height	7																		
Tibia height	8																		
Sitting height	9																		
Nape to waist	10																		

■ *Indicates a high correlation of over 0.7*

▦ *Indicates an appreciable correlation of 0.5 or over*

Figure 47
The correlation matrix (Table 12b) is here diagrammatically summarised distinguishing high and appreciable associations. Two strong groupings emerge. From Croney, J (1977), *Man* (NS) 12

although the total variation from which they were calculated may be different.

The preparation of a correlation matrix (Table 12b) and then its transformation into a matrix where groupings are defined by their degree of correlation (Figure 47) could be important in the early design stage because the grouped values afford easy graphic reference to high, appreciable and low associations between variables. This is very useful if one has to go on to determine the proportion and shape of a clothing or ergonomic 'envelope' or 'tent' for easy usage, because an optimum 'envelope' is developed from a designer's awareness of human dimension relationships. In this particular example two strong groupings emerge; weight and girths together with the horizontal measurements of bitrochanteric width; and secondly total height and all other indirect linear measurements of body segments. Arbitrary levels have been selected for this figure, but for the purpose just mentioned this does not detract from its value. Above a 0.5 correlation value indicates a 75% chance or more of predicting one measurement from another in the relevant grouping, 0.7 indicates an 85% chance or more for prediction. Figure 48 is a further example from a larger survey.

Having calculated and displayed various values of r it is nearly

MEASUREMENTS GROUPED BY SIMILARITY AND SIZE OF CORRELATIONS

		1	14	15	17	19	16	9	13	28	7	20	8	35	23	27	36	18	11	12	37	2	3	30	4	5	34	31	22	6	10	21	24	25	26	29	32	33
1	Weight		X	X	X	X	X	X	X	X	X	X	X	X	X	O	O	X	O	O	O									O								
14	Bust girth	X		X	X	X	X	X	X	X	X	X	X	O	X	X	O	O	O	O	O																	
15	Waist girth	X	X		X	X	X	X	X	X	X	X	X	O	X	O	O	O	O	O	O																	
17	Hip girth	X	X	X		X	X	X	X	X	X	X	X	X	O	O	O	O	O	O	O																	
19	Chest girth at scye	X	X	X	X		X	X	X	X	X	X	X	O	O	O	O	X	O	O	O									O								
16	Abdominal ext. girth	X	X	X	X	X		X	X	X	X	X	O	O	O	O	O	O	O	O	O																	
9	Abdomen-seat diameter	X	X	X	X	X	X		X	X	X	O	X	O	O	O	O	O		O	O																	
13	Bust arc anterior	X	X	X	X	X	X	X		X	O	X	O	O	O	X	O	O	O	O																		
28	Upper arm girth	X	X	X	X	X	X	X	X		X	X	O	O	O	O	O	O	O	O	O																	
7	Scye width	X	X	X	X	X	X	X	O	X		X	O	O	O	O	O	O	O	O	O																	
20	Armscye girth	X	X	X	X	X	X	O	X	X	X		O	O	O	O	O	O	O	O	O																	
8	Bitrochanteric width	X	X	X	X	X	X	X	O	O	O	O		O	O	O	O		O	O	O																	
35	Knee girth	X	O	O	X	O	O	O	O	O	O	O	O		X																							
23	Neck to bust	O	X	X	O	O	O	O	O	O	O	O	O			O																						
27	Width of bust prom.	O	X	O	O	O	O	X	O	O	O	O	O		O				O																			
36	Calf girth	X	O	O	O	O	O	O	O	O	O	O	O	X	O																							
18	Across back	O	O	O	O	X	O	O	O	O	O	O																		O								
11	Neck base girth	O	O	O	O	O		O	O	O	O							O																				
12	Across chest	O	O	O	O	O	O	O	O	O	O	O						O																				
37	Body rise	O	O	O	O	O	O		O		O		O																									
2	Stature																						X	X	X	X	X	X	O									
3	Cervical height																					X		X	X	X	X	X	O									
30	Arm length posterior																					X	X		X	X	X	X										
4	Hip height																					X	X	X		X	X	O										
5	Tibiale height																					X	X	X	X		X	O										
34	Side seam																					X	X	X	X	X		O										
31	Arm length anterior																					X	X	X	O	O	O											
22	Nape to waist																					O	O															O
6	Interacromion width	O											O							O																		O
10	Shoulder slope																																					
21	Depth of scye																																					
24	Centre shoulder to waist																																	X	X			
25	Anterior waist length																																X		X			
26	Cervical to front waist																																X	X				
29	Shoulder length																																O					
32	Trunk line																												O									
33	Waist to hip																																					

x—Correlation of 0·7 and over

o—Correlation of 0·5 but under 0·7

Figure 48
This is another example showing that correlation values of body measurements form associated sets; and that dimensions can be taken from the two major sets, weight and girths, or stature and segmental lengths, to act as 'key' measurements or representative dimensions that explain a large part of the association. From Kemsley, W F F (1957) *Women's Measurements and Sizes*, HMSO

always necessary to determine by how much an increase or decrease in one body dimension may be expressed as a unit increase of another variable. When the increase of a variable is predictably regular in relation to another variable, one dimension may be estimated from another from a point on a line drawn to accommodate them. (Figure 49). These straight lines are regression lines, and providing there is an intense association between two variables, the estimation of one value from another can be quite accurate.

It may be realised from looking at the matrix of grouped r values (Figure 47) that regression analysis will show the relation-

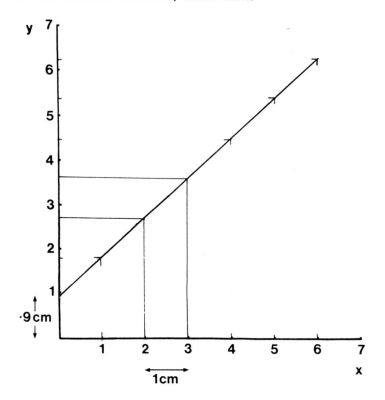

Figure 49
Regression line for women's cervical height on their total stature. The increase of cervical height, the dependent variable, is predictably regular in relation to total stature, the independent variable. For every increase of 1 cm of stature there is an increase of about 0.9 cm in cervical height

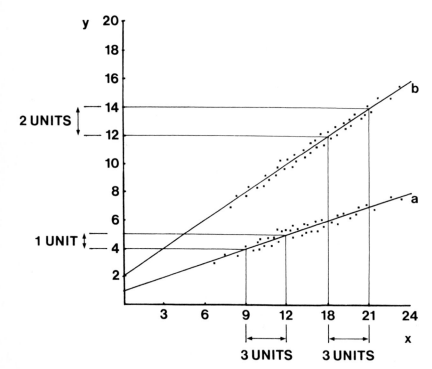

Figure 50
Two trend lines (**a** and **b**) for different bi-variate distributions. The slope of trend line **a** represents data where for every change of 3 units in the 'key' measurement there is a change of 1 unit in the dependent variable. The slope, and therefore the length of the axial intercept, of trend line **b** represents different data where for every change of 3 units in the 'key' measurement there is a change of 2 units in the dependent variable

ship in the two main groupings to be linear or very nearly so. It must not be assumed from this that human body measurements fall only into two groups; 'there remain however a number of body measurement variables such as nape to waist and inter-acromion width that do not readily intercorrelate with the two strong groupings of body measurements; namely of bulk (weights and girths), and of stature and associated linear measurements. They may well form a third grouping of their own (Kemsley 1957). They comprise sub-divisions of trunk and limb measurements which are frequently used in the clothing industry. They may be defined as measurements which follow body contours, but which are not always complete girth circum-ferences or segmental lengths, and which are not necessarily parallel to the three main anatomical planes (Croney 1977)'. The relationship in this third group may not be linear. Nor may there be a linear relationship between a variable from this third grouping, if it is a cohesive group, and the other two groups. This remains a matter for investigation.

From what has been stated so far it should not be thought that regression lines can be guessed from a bi-variate distribu-tion; scales and slopes of lines must be calculated from a precise point.

What is required is the location of a point on line of 'best-fit', a trend line which cuts all the points where the two variables are paired. From the first point we need to know how much of measurement represented by the line (its slope and its length which gives the axial intercept) is changed by the change in the 'key measurement' (Figure 50); or how the dependent variable changes in relation to a unit change in value of an independent variable. 'Key measurement' is a practical way for a designer to think about independent variables, because nearly always a small number of body measurement variables are the 'keys' to the anthropometric solution of ergonomic problems. A func-tional relationship between the 'key' variable and a dependent variable starting from their origins (0, 0) is not known with anthropometric variables.

The equation $y = a + bx$ gives the mathematical form for the relationship of y and x if it is a straight line, y being the value of the dependent variable and x the value of the independent variable; the determined constant a is the expression of the slope or angle of the regression or 'trend' line the determined constant b the y intercept value, or the point at which the y axis is cut by a projected line. With a straight line a and b values can always be found. The problem is that two similar equations with different values for a and b can give nearly the same value of 'best-fit'. What has to be done is to find an equation value that gives the smallest residual variance in the y ordinate values. This could be done by trial and error, but is better referred to the differential calculus method of solution invented to resolve this problem.

The computation of regression lines for 'best fit' not only requires the calculation of the variance of the two distributions x and y, so the calculation of $^s x$ and $^s y$, but the calculation of the estimate of covariance written $\sum dx\, dy$.

The complete equation for y on x looks as follows:

$$y - \bar{y} = \frac{\sum dx\, dy}{\sum d_x^2}\, (x - \bar{x}) = \frac{^s xy}{s_x^2}\, (x - \bar{x})$$

This equation is based on ideas first formulated by Gauss and uses the Principle of Least Squares taking squared deviations for 'best fit', d here standing for a deviation or difference, with or without sign.

s is the estimate of standard deviation, s^2 the estimate of variance, s_x^2 the estimate of the variance of x. In statistics the Greek letter is used for a parameter and a equivalent Latin letter for an estimate. σ = standard deviation, s is the estimate σ.

It now can be seen that regression calculations are bulky and much complication can be generated on the subject. It is suggested that the designer should seek professional assistance with these calculations, not only because the calculations are arduous, but for, as has already been said, large amounts of data can now be quickly dealt with by mechanical calculations.

As estimates of one value from another can be accurately predicted from regression lines, it is not entirely necessary for every body dimension to be taken on a sample of users for a new design or ergonomic system.

The statistical analysis of anthropometric data is an attempt to represent the dimension of users. Much is done by the experienced judgement of the designer, the rest is done by the use of typical measurements found to be reasonably common in the majority; or alternatively the abstraction of a measurement limit, or relationship of a limits, that will serve a good percentage of people. It is necessary quite often to add tolerances to measurement dimensions. These are added by experienced judgement for large user populations; for smaller populations they are extremely critical and can be examined by models rather than statistical analysis.

Conclusion

When data is organised into tables or graphs it must not be too overcrowded. Other data from previous surveys for the same object should be organised in an equally clear and simple manner so that the whole can be compared. This editing of earlier work of a similar nature should be undertaken before the new survey is started. Measurements from a previous survey will assist in decisions about relevant measurements for a new survey.

All anthropometric sampling should be considered as research leading to the positioning or clothing of a person in one of two situations; because all conceivable environments, when considered in relation to the possible physical activity of its users, fit into one or the other of two main groupings.

One grouping will include all those objects or coverings of everyday use and wear. This grouping covers all access ways and spaces, furniture and architectural features, dimensions and clothing, etc.

The other grouping will include only equipment and clothing for highly specialised use and needed by a minority of a population. This grouping covers all special operating situations in industrial or commercial enterprises, in communications and endeavours in human aid.

It follows that for general use any design which imposes limitations by using unsuitable dimensions unnecessarily complicates its use.

Indices

When we examine the shapes and proportions of the figures of a population we find that their characteristics can be more easily conveyed by combinations of measurements than by single measurements. These combinations we call indices. Indices state in the form of a percentage the relationship between two measurements. Indices are determined by multiplying one measurement of a bodily dimension by 100 and dividing the result by the measurement of another dimension. For example, an index which has been very widely used in physical anthropology is the cephalic index. It gives the relationship between head length and head breadth:

$$\text{The cephalic index} = \frac{\text{Max breadth of skull} \times 100}{\text{Max length of skull}}$$

If the result gives a figure below 75% the head is said to be dolichocephalic

> a figure between 75—80% mesocephalic
> a figure above 80% brachycephalic

Dolichocephalic is a type of head that is longer than it is broad. Mesocephalic a broader headed type. Brachycephalic a broadness of head which is verging on roundness.

Two other examples follow:

The cormic index gives the relationship between standing height and sitting height.

$$\text{The cormic index} = \frac{\text{seated height} \times 100.}{\text{standing height}}$$

An answer of 45% would indicate a short trunk to leg length and of 55% a longer trunk to relatively shorter legs.

The thoracic index gives the relationship between two

dimensions of the thorax. It varies naturally with body build, but more especially with sex and age.

$$\text{The thoracic index} = \frac{\text{lateral diameter of thorax} \times 100}{\text{anterior-posterior diameter of thorax}}$$

The adult thoracic index ranges between 130% and 135%.
Some more indices are:

Ponderal index: stature divided by the cube root of weight.
Bust index: chest depth/chest breadth.
Waist index: waist depth/waist breadth.
Shoulder index: bustpoint-to-bustpoint breadth/biacromial breadth.
Hip-shoulder index: hip breadth/biacromial breadth.
Hand index: hand breadth/hand length.
Lower limb girth index: calf circumference, right/upper thigh circumference.
Foot index: foot breadth/foot length.

A large number of other indices exist but it is not necessary to list them all here because they have only the most marginal application to the design process. Once their function of imparting an idea of relative proportions is understood they can readily be invented when necessary.

Constitutional indices

Many attempts have been made to contrive indices that will indicate levels of bodily health and composition. They have not proved to be entirely adequate or reliable. But when one considers the fluctuating ranges of the many parts of man's physiological system this is not to be wondered at. To compile a constitutional index is like attempting to get a quart into a pint pot. The following is an example of a constitutional index to indicate underweight or overweight. The inventor was the German anthropologist F H Lorentz.

Age in years

2.5 to 6	Height-weight-(height-123) $\times$ 0.7	= 100
6 to 14	Height-weight-(height-125) $\times$ 0.5	= 100
14 to 18	Height-(weight-10)	= 100
18 and above	Height-weight-(height-150) $\times$ 0.25	= 100

Where the height is in centimetres and the weight in kilograms

Over 100 indicates underweight

Under 100 indicates overweight

5 Dynamic anthropometry

Introduction

When undertaking the design of a piece of equipment or furniture or any environment for work or leisure we are always seeking an ideal or optimum situation for a person or persons. The ideal or optimum will be relative to a particular situation because, for example, what is 'comfortable' for leisure sitting, may not be seen in quite the same way when we are looking for a comfortable seated position which will at the same time allow maximum of effort to be exerted by a limb for a certain period of time, with the minimum of fatigue. So optimum for comfort might not always be able to be arranged to coincide with optimum for effort. What we can hope to do is so to design equipment, or space, that people using it are as well suited as the situation will permit and their tasks demand. This requires a great deal of imagination in the use of facts and figures about people. Man has limitations and many possibilities: not least in the last respect, an ability to stand up to stress. The designer must know that there is a price every user has to pay eventually for going beyond his capacity. The body is made of deformable materials, and man has very little instinctive ability left to tell him when his physiological processes have had enough. Undue stress on the body must therefore be curtailed. There is, as always, the other side of the coin. Too much rest and comfort lead to boredom. So that although undue physical trial should be eliminated, some evaluated system needing manipulation and selection should be left to enable the brain to be exercised. Man at his best can make new relationships visually or in the form of mental calculations, invent new combinations of control, initiate new actions, and react to slight changes in range sensory stimuli. He can also perform very delicate adjustments. These things may still be left to him, and other tasks may be passed onto machinery.

The construction of manikins can be recommended for the initial stages of an industrial or architectural design. If they are accurately measured and constructed they give the facility of an efficiently presented set of inter-related measurements. They can be moved easily over the drawing board and can supply some speedy answers. However, they can only give information about a problem from a rigid, two-dimensional point of view. The designer must take his enquiries a great deal further if he is to satisfy the requirements of a live operator making dynamic movement in three dimensions.

During the initial design stage as much use as possible should be made of all relative and existing data. Engineers' and other designers' findings are relevant, as are comments from users and manufacturers. A continual return to the drawing board should be made to set each piece of new information into its place alongside the original draft. Other solutions of a similar design-problem should also be seen, and they should be probed for any weaknesses.

With a comprehensive set of data that will accommodate a normal range of operations set out in the designer's drawings, trials must then be carried out. This necessitates a 'mock-up' being constructed of the design. This must be considered a very vital part of a designer's job. If in the trials people have to be used to simulate operators or users great care should be taken that their physique is typical of the population of users of the situation one is constructing.

Trials bring to the forefront detailed consideration of allowances for ease of movement, vision or operation and they will always pinpoint the special conditions of particular cases. The extremes of the range of users should be carefully examined against specific operations.

Designing for human users is largely problem-solving in a physical environment. By trials we can design the environment and finally mould it successfully into an optimum envelope for use. The dynamic movements of a typical group of real users give final meaning or the lie to sets of anthropometric measurements. Imaginative and theoretically correct solutions can be vindicated by trials. The trial stage itself must be conducted rigorously by the designer to discover any weaknesses that are dormant. Improbable, but possible, combinations of movement and use must be examined. As improbable combinations of movements can often be the cause of accidents it is of great importance that they should be studied and if possible be eliminated. Their probability and the extent to which they can weaken a system should be probed with practical tests.

If there are problems of use against time or force or of accuracy, recorded observations will suggest necessary alterations. If pre-packaged equipment must be introduced into a new situation the dimensions of movement and force required to operate it must be carefully examined so that it can be 'harmonised' with other equipment about it. It may have to be given pride of place in the user's cone of vision or easy range of reach.

The introduction of pre-packaged equipment brings to the forefront the problems of maintenance access. If maintenance is to be carried out by the user it needs special consideration in the initial design stage. Any form of maintenance of equipment, however, should be considered when a list of requirements for the new design are being undertaken.

Ergonomic factors in design planning

Study the purpose and the operating performance details of the equipment, furniture, appliance, mechanical device or housing. Study the design solutions already in existence.

Dimensions and space

Allow for clothing, access, and movements in and about the work space. Attempt to accommodate between 5th and 95th percentile of users, or alternatively specific users. Consider any maintenance problems.

Operating conditions

Wherever possible use adjustable equipment. If users cannot be specified planning should consider women's percentile measurements rather than men's. If design is for non-specified users and there are any possible side effects from use, or difficulties of operation, they should be reduced to a level that could be tolerated by the older age range.

To facilitate speed

Consider the correct grouping and positioning of control, and the grouping of indicators within the normal cone of vision. For average use limits on distance will be set by the lesser body measurements.

To facilitate precision

Consider support for body or limb segments and the operating position.

To facilitate pressure or force

Consider the amount of pressure or force and whether whole or part of a limb must be used. Consider body or limb supports and space necessary for operator to make adjustments of position.

To offset fatigue

Consider grouping of controls and the operating position in detail.

Trials

Make scale models and test. Use a full scale 'mock-up' of design. Check that the trial subjects are adequately representative of the future users; particularly that they represent the low and high percentile values of users. Check all detailing thoroughly. Determine all the limiting factors of the design, and ascertain by experiment whether they detract from optimum performance.

Adaptation

Man reacts to his surroundings. His surroundings can affect his whole performance.

A number of changes in his environment can be tolerated. It is possible to adapt to higher temperatures than one is normally accustomed to. It is also possible to continue working when the average light intensity is decreased. In more special fields of human endeavour it is possible to adapt to a lower air pressure and also to travel at faster speeds. That is not to say that if these changes are major that sudden adaptation can take place. Most new design situations demand some process of adaptation however slight, and generally they are very readily tolerated. There are two changes in a person's surroundings that are harmful to performance and cannot be tolerated: they are changes towards brighter lighting and louder noise. Both of these are equally distressing and prove extremely difficult, if not impossible, to adapt to.

Anatomical movement analysis

The human skeleton consists of rather more than 200 bones. (Figure 51.) Bone is an extremely tough and hard substance, but it does exhibit remarkable qualities of plasticity. It is being continually remodelled in the process of growth, and makes a contribution of its own to the physiological processes of the body. However, it is the supporting framework for the muscles and their actions providing bases and pillars which can resist many combination of stress. This second duty demands that the bones should be strong but not brittle, and that they should be resilient enough to be able to withstand quite large forces of stress and strain from an almost limitless series of directions.

The evolution of the body's framework of bones has given the possibility of many combinations of movement, and it is unique amongst animals in the types of joints it can offer. By the co-ordinated use of muscles and ligaments a large part of the framework can be held rigid, whilst a small portion is in action. To alter the position of the body as a whole or in part, the bones of the appendicular skeleton also offer a well-planned system of levers. The human body is a structurally adaptable instrument which, in normal health, can execute the dictates of man's will and conduct the most complex movements with the minimum of conscious control. We have only to consider the series of combined movements needed in an everyday occurrence like walking, and simultaneously rearranging objects being carried, to understand this.

In the process of structural adaptation the architecture of bones has somehow been able to make use of civil engineering skills of a very high order. If we examine the femur bone of the thigh we can see that it has been tooled into a large number of subtly curved surfaces; and without any mathematical analysis we instinctively realise that the many facets of its surface reflect its mechanical purpose in locomotion and our upright stance. Like the femur, each bone of the skeleton is designed for a specific number of roles.

Man's skeleton shape is such that he can manage short periods of normal standing posture without much assistance from his postural muscles, and in this position he is relatively efficient. However, the state of equilibrium is an uneasy one. After standing for some time the weight of the upper body does threaten the knees with collapse, and when standing for lengthy periods we tense the muscles about the leg joints to resist the pressure from above. These short bursts of muscle action mean

Figure 51

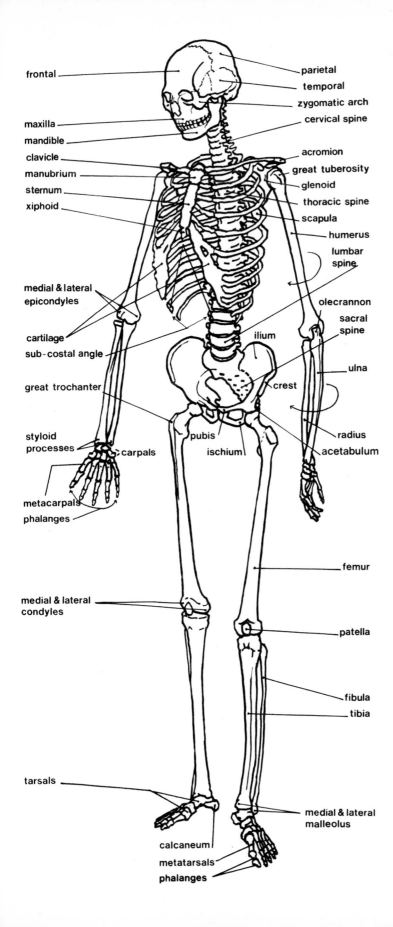

frontal

parietal

temporal

zygomatic arch

cervical spine

maxilla

mandible

clavicle

manubrium

sternum

xiphoid

acromion

great tuberosity

glenoid

thoracic spine

scapula

humerus

lumbar spine

medial & lateral epicondyles

cartilage

sub-costal angle

great trochanter

olecrannon

sacral spine

ilium

crest

ulna

styloid processes

carpals

pubis

ischium

radius

acetabulum

metacarpals

phalanges

femur

medial & lateral condyles

patella

fibula

tibia

tarsals

medial & lateral malleolus

calcaneum

metatarsals

phalanges

that standing for long periods necessitates the use of energy.

In assuming the upright stance man disregarded the advantages the quadruped has in the positioning of his thorax between four supports: the spinal column in this position acting as a true cantilever type bridge between the fore and aft of the animal. Man's thorax hangs from the spinal column. To solve the problem to some extent man has a shortened lumbar spine and an enlarged sacral spine. This allows a fair amount of turn at the waist without too much loss of stability. (Figure 51.) All the bony formations are bound together by the more elastic connective tissues of joint capsules and ligaments. Establishing a preliminary classification for skeletal study, we designate the thorax and the skull the axial skeleton, and the limbs the appendicular skeleton. The thorax and skull are joined together by the vertebral column which, by the nature of its construction of 29

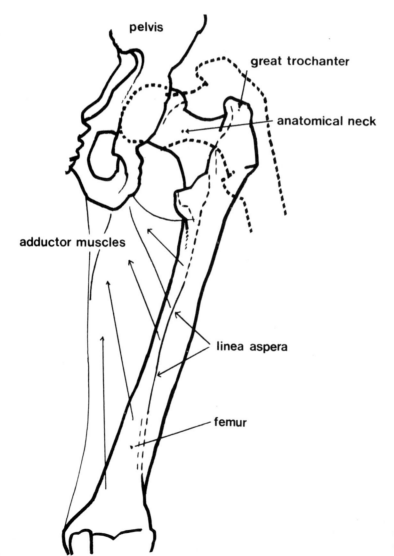

pelvis

great trochanter

anatomical neck

adductor muscles

linea aspera

femur

Figure 52
Diagram of the femur bone, posterior view. The linea aspera is a raised ridge on the bone. The dotted line shows the head of the femur bone in abduction

separate bones, excluding the coccyx, provides a flexible link. The vertebral column as a whole consists of five distinct sections. The cervical spine or neck (seven pieces), the thoracic spine (twelve pieces) relating to the chest cavity, the lumbar spine (five pieces) the sacral spine (five pieces) which is permanently joined and is the vertebral column's base in the pelvis, and the coccyx which variously contains three or four, or even more rarely five pieces (61% of people have four pieces). The column allows us particular freedoms of movement. The cervical spine allows up to 90° of movement in bending and the lumbar spine up to 30° of movement in bending, the cervical spine allows rotation on its superior end of up to 55° rotation from the medial line, and the seven bones in combination allow 40° unilateral bending and between 45° and 50° in flexion and hyperextension in the sagittal plane. The thoracic and sacral spines are curved posteriorly and are present at birth. The cervical spine is convex-anteriorly and is shaped when the 6 to 8 months old human sits up; the lumbar spine is also convex anteriorly but is shaped slightly later at about 12 to 14 months when walking has been achieved.

The skull is considered anatomically in two parts; the bones of the cranium and the bones of the face. The cranium contains and protects the brain, the face contains the organs of recognition and communication and the palate.

The thorax contains and protects the lungs and the heart. Man's thorax is proportionally broader than that of other animals and it is also somewhat flattened in its anterior and posterior aspects. The greatest transverse section is in the lower half below the xiphoid process. The upper section is tapering in construction, which allows space for the movement for the four joints at the shoulder girdle from which the arm springs. (Figure 53.) Although rigid enough to support arm movements the bones of the thorax cage move in respiration.

The thoracic cage moves anteriorly, posteriorly and laterally independently of the pelvis in the region of the lumbar spine and can in this way improve average reach distances by up to 406 mm (16 in.).

The bones of the pelvis form an open rimmed bowl-like structure. It supports the abdominal contents with laterally-facing bony plates. These bony plates, or hip bones, are joined together at the back by the sacrum and in front at the symphysis of the pubic arch, and they provide firm socket positions into which the head of the femur bone can be joined to form a secure ball and socket joint. (Figure 52.)

During walking upper-body weight is transmitted through the pelvis to the legs. The pelvis is built for strength and in itself has a very limited range of movement. A joint between the sacrum and the ilium, the sacro-iliac joint, allows necessary movement between the base of the spine and the pelvis. By means of this joint primary movements can be made with the pelvis to re-adjust

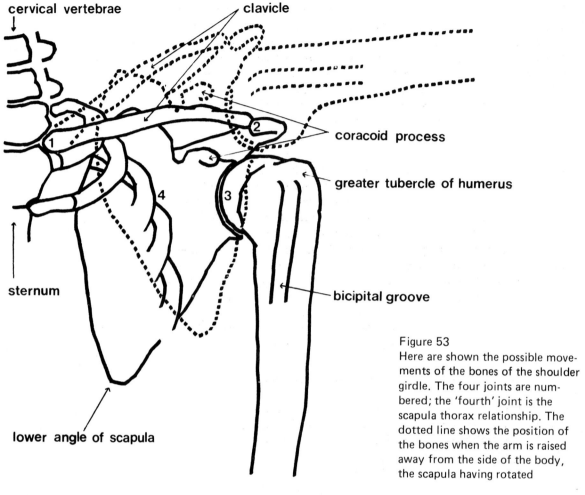

cervical vertebrae

clavicle

coracoid process

greater tubercle of humerus

bicipital groove

sternum

lower angle of scapula

Figure 53
Here are shown the possible movements of the bones of the shoulder
girdle. The four joints are num-
bered; the 'fourth' joint is the
scapula thorax relationship. The
dotted line shows the position of
the bones when the arm is raised
away from the side of the body,
the scapula having rotated

its tilt. This is necessary in the re-alignment of body parts in
postural changes and in the use of the legs in activities other
than walking. Weight from above spreads it out laterally. The
pelvic bowl in its external aspects affords mechanically advantageous anchor points for muscles in their action on the thigh
bones in walking and running; the pelvis is much broader in man
than other animals for this reason. The whole trunk can be
assisted to bend forwards when the legs are fixed by the flexor
muscles of the thigh pulling from the front surfaces of the
femur. Conversely the extensor muscles of the thigh can be used
to erect the trunk. Many of the thigh muscles are two-joint
muscles which means that because they cross two joints they
can be seen to have preferred ends of action; acting proximally
they flex or extend the trunk on the legs, acting distally they
flex or extend the lower leg on the thigh. A number of muscles
associated with bone movements at joints in other body regions
may be seen to act in a similar manner.

The legs, commencing from the acetabulum socket in the
pelvis, consist of one long upper bone, the femur, and two

lower long bones the tibia and fibular, and then a number of smaller bones which make up the foot. The femoral-tiboral joint is capped by a small separate piece of bone, the patella. The femur bone has a unique and lengthy anatomical neck which facilitates the anterior-posterior movement of the bone by keeping it clear of the pelvic sides, and allows its greater trochanter to be an unobstructed anchorage point, almost in the form of a lever handle, to the muscles which move it. With all tissue removed, the femur direction gives the human skeleton a marked bandy appearance. But it is this inward movement of the femur bones, coming together at the knees, which meets and contains the downward thrust of the mass of the trunk and skull about the centre of gravity. Strong adductor muscles pulling towards the pubic arch have the almost impossible task of preventing the lower appendages from splaying out. (Figure 52.)

The somewhat bandy appearance of human legs is heightened in the female figure, where with rather shorter femur bones and a relatively wider pelvis, the femur pelvic triangle comes slightly nearer an equilateral triangle. This is a prominent secondary sexual characteristic which affects a woman's gait. Because of the equilateral relationship there is a more noticeable dip and rotation of the pelvis during walking.

The legs are not only capable of anterior and posterior movement, but can also be pulled laterally away from the central body line in abduction and returned in adduction. This, in practice, allows each leg complete cyclic movement when the mass of the body is balanced on the non-preferred leg.

The tibia with the fibula complete the other half of the linear height of the legs, and supply strong and nearly perpendicular terminal posts onto the arch of the feet. The whole leg length can be rotated away from the medial line of the body from the ball and socket joint at the hip. This is a combined movement which with normal muscle action has a 'follow through effect' in the leg and finally effects a lifting action on the arch of the foot. The length of man's legs are necessary to allow for the locomotion movements he wants. This in turn demands powerful muscles; and the well-marked ridge on the femur bone, the linea aspera, which is the insertion of the adductor muscles of the thigh, gives a good indication of the power of the leg muscles by its prominence.

Man's foot is peculiarly his own. He has adopted the plantigrade position and walks on his extended foot. A number of bones of varying size form the foot by a series of arch like forms lying side by side. The arch of the foot, which is maintained by ligaments across the bones, provides an elastic spring board for the whole body mass when walking. (Figure 54.)

If the motion of walking is considered the conclusion would be reached that there is not just one kind of gait that should be considered normal, because different somatotypes give a different appearance to an individual's movement.

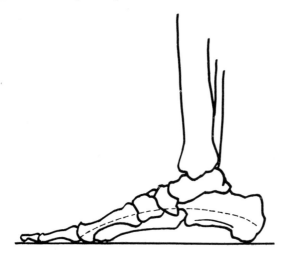

Figure 54
Diagram of the bones of the foot
to show their arch-like form. This
diagram shows the medial side

The head moves up and down in walking, and we are less tall when in motion. The centre of gravity is lower during motion and also shifts sideways from one leg to the other. Body motion requires an inter-play between the loss and recovery of balance. Each stride forward is a pursuit of a newly placed centre of gravity, each walking cycle going from heel strike to heel strike. From the first heel strike body weight is moved across the stance foot from the heel to the ball of the foot. As the heel of the stance foot rises the swing foot is making its heel strike and so on.

The arm consists in its upper segment of one long bone, the humerus, in the lower segment two long bones, the ulna and radius, and in the most distal segment of all a number of bones form the hand. The arms, by the nature and relationship of their joints, have a much freer range of movement than the legs.

The arms and the hands together are motivated grabs which can be extended into an almost undefinable series of positions in relation to the medial line of the body. The arm in total can be passed behind the head, it can also be passed diagonally down behind the trunk; it can also move out in several directions away from and across the front of the trunk. To shorten these distances or to lift or lever the arm can be flexed half way at the elbow. The hand can be rotated into varying positions by the radius bone being passed over the ulna at the wrist end. This action is called pronation; the equal and opposite return movement is called supination. (Figures 55 and 56.) Varying degrees of pronation and supination can be executed even though the long bones of the arm are presented in a variety of positions to the trunk. Finally the hand itself provides a claw like grip.

This extraordinary range of possible movement stems from the fact that the arm is not fixed to the trunk by one joint, but by four joints. The four joints, progressing from the medial line of the body are: posteriorly, the 'scapula-throacic' not a true joint, but the movement of the scapula over the posterior thoracic wall is necessary in most movements of the shoulder

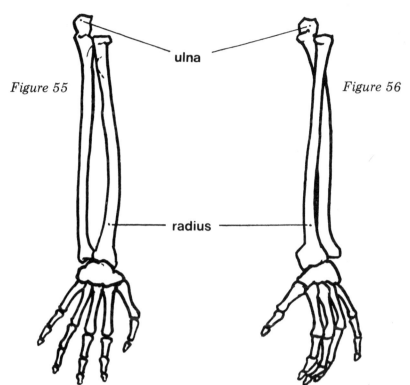

Figure 55

ulna

radius

Figure 56

Figures 55 and 56
Diagrams to show the meaning of
supination and pronation of the
forearm

55
This is a front view of the bones of
the left forearm and hand. The arm
is in a supinated position with the
thumb facing laterally away from
the body. The radius and ulna lie
side by side.

56
The arm remains in the same
position, but the distal end of the
radius bone has turned over the
ulna twisting the hand at the wrist
so that the thumb faces medially.
This is the pronated position of the
forearm, and the radius head has
been rotated to execute the movement

The movement of pronation
and supination is seemingly
increased when the elbow is fully
extended because of the addition of
some rotation by the humerus bone

girdle (4); anteriorly, sternoclavicular (1); then laterally, acromio clavicular (2); and finally the scapula-humerus or shoulder joint (3). (Figure 53.) The scapula-humerus joint is a semi ball-and-socket joint. It acts in the manner of a crane-head in all the lifting operations of the human arm, the pivot base of the crane gantry being formed by the scapula-thoracic and the clavicular-sternum joints. The main sideways hoist or shrugging movement with the arms commences at these two joints. The clavicular-scapular joint enables an easement to take place in the shoulder girdle, in the position of the crane head, and may offset a possible breakage in what might otherwise be a too rigid system. Altogether the bones of the arm and the hand offer a very flexible series of bony links.

Joints
A joint is the formation, or bringing together, of any two skeletal parts by means of another more pliable structure which allows movement. (Figure 57.) The amount of movement at a joint depends largely on the type of surface the two skeletal parts present to one another, and the manner in which they are held together. The type of joint and its position in the skeleton can be related to the demands that are placed upon it. The different types of joints are undoubtedly descendants of earlier varieties which have been modified in the course of time. Contemporary man now has a useful mechanical system though no precise study has ever recorded its many sided capabilities in

terms of efficiency. Other animal systems have confined themselves to a more predominant use of the simple ball and socket joint. Human development has expressed a preference for uniaxial joints[1], and it has been further suggested that the selection of this type of joint was evolved because it makes the most economical use of muscular energy. Therefore in any deliberations that are concerned with the human operator and his positioning in relation to controls consideration must be given to the most advantageous use that can be made of the linked flexing extending movement he offers. Two bones in a joint have only small portions of their surfaces in contact at one time, the greatest contact often being found at the limit of a joint movement.

At any position of possible joint movement contact is maintained between the remaining surfaces of bone by smooth cartilage discs. The bone ends are also capped with hyaline cartilage, and the joint is securely bound together with ligaments. The cartilage acts as a buffer, and, because of the elasticity of the cartilage, the fit of the joint is maintained during movement. Degree of fit or congruence is related to the type of construction (bone and joint capsule) of a particular joint. The cartilage buffer shields the bone ends from wear, and with the maintenance in a healthy person of the self-lubricating system that every joint has, friction is of a low order.

Human joints can either be classified by movement or by the character of the tissues of the union. Classification by movement is the most convenient when studying the type and range of action of an operator's body or limbs. Classification in this manner gives us three types of joint: Immoveable joints, or the joining of two bones which become permanent during the processes of growth, and where no movement can take place. Slightly moveable joints, where a limited movement is allowed. Freely moveable joints, where extensive movement is possible. It is the last two types of joint that merit a designer's attention; and the last type of joint in particular which covers all major executive movements in the limbs. The very free movement of the shoulder joint needs particular consideration by the clothing designer.

The joints of free movement can allow bones to take up varying positions in relation to one another, in all three dimensions. Some joints however only allow movement in one dimension. With all the free joints it is not possible to define precisely their arcs of movement. Distal ends of bones in the appendicular skeleton can point in an infinite number of directions. Every total movement of a limb is an additional sum of the anatomical movement of adduction, abduction, rotation, flexion and extension of the joints of that limb. These move-

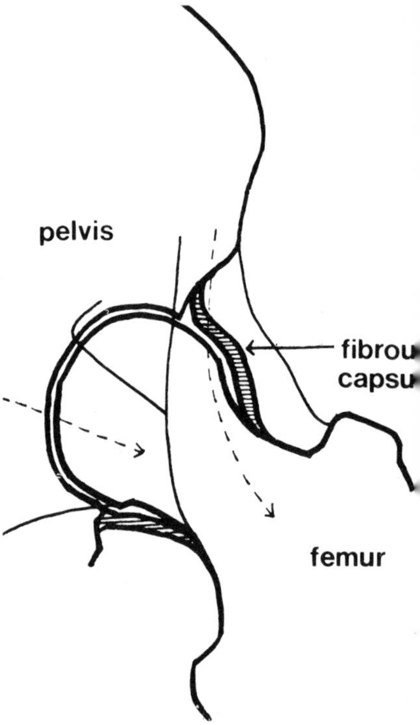

Figure 57
A diagram to show the simplified version of a joint. The heavy black lines show the two bony parts in cross section.

The fibrous capsule surrounds the joint completely keeping the bones together. The dotted lines represent the direction of bands of ligaments, which are thickened outer portions of the capsule, the whole binding the bones together with great tension and strength

[1] Barnett, G H, Davies, D V and MacConaill, M A, *Synovial Joints*, 1961. Section 2, Chapter 6.

ments are explained in Figures 63, 64, 65 and 66, the range of movement of a limb being increased or decreased by the position of the trunk.

Healthy joints are so manufactured that it is exceedingly difficult in normal usage to overcome their limitations of movement and damage them. Joints can resist a sudden force as well as a continuous one.

Arcs of movement in all three dimensions depend on bone construction, associated muscles and ligaments. Joints in the appendicular skeleton can become stationary by the checking action of the attached muscles and ligaments. Conversely, some rotation can take place at a number of limb joints. Alternatively, rotation of the distal end of a limb can be increased by rotation at the proximal end of limb at the trunk. For example the action of the deep buttock muscles rotates the femur bone, which movement is transmitted right through to the foot, assisting it to conduct a lateral rotating movement. Alternatively, when the tibia and femur bone are at right angles, the tibia bone is able to rotate to some degree with the foot.

The arm possesses much free movement. (Figure 58.) In certain arm movements its swing takes it through three dimensions of movement simultaneously, in other movements different dimensions of movement have to be achieved independently. For example when the arm is hanging by the side of the body palm inwards, to achieve the abduction of the arm to a level of the shoulders with the palm of the hand facing forwards two

Figure 58
Diagram to show movements of the joints of the arm. The joints of the leg have the same series of movements; multiaxial at the thigh, biaxial at the knee and uniaxial in the foot. The great difference between the hand and the foot are that the phalanges can be opposed across the palm of the hand, whereas the phalanges of the foot make a series of parallel arches

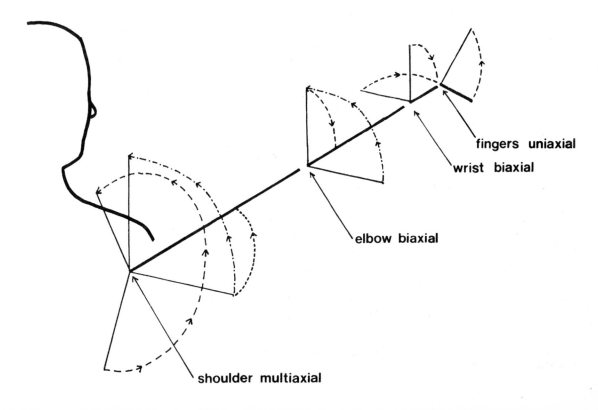

fingers uniaxial

wrist biaxial

elbow biaxial

shoulder multiaxial

distinct movements are necessary: arm abduction and wrist supination. The same final hand position can be acheived by an arm swing in different spatial dimensions; first the whole arm is flexed at the shoulder and then is drawn back to the shoulder line by horizontal abduction. By the second means humerus rotation and wrist supination go unnoticed. Similar experiments can be conducted with the leg.

The arm fully flexed at the shoulder offers the operator a more deliberate torque at the wrist. It can well be used in the standing position where reach may be extended by body movement. When the arm is raised at the shoulder by flexion shoulder and wrist action combine well together to facilitate deliberate hand use on large scale. Nevertheless these predominately shoulder movements will require practice on the part of an operator.

With the elbow in a partially flexed position an operator can conduct slight pronating and supinating movements for calibrated dial adjustment with near viewing. Because in this position the elbow and wrist joint movements may be combined to offer the widest range of movements for the hand. Most joint movements are of a conscious nature in the conception of an action, a vital proportion are actionable by conditioned reflexes and establish everyday positions the body requires. Quick movements to escape damage to vital parts come into the same category.

Muscles

Muscles work by contraction. They consist of bundles of cell-fibres. A muscle fibre can shorten its length by at least one third in contraction. (Figure 59.) Contraction brings the two ends of a muscle nearer together. Therefore it brings the two bones to which it is fastened nearer together. Muscles as contractors always pull. Even when, for example, we push down with our forearm it is because of the pulling action of the triceps in the upper arm. The energy of a muscle is expressed in contraction against weight or resistance and is the result of bio-chemical-physical processes. These processes are advanced by impulses along the motor neuron of the nervous system from the brain to the spinal cord. Conduction of impulses passes from the brain to skeletal muscles to determine voluntary actions. Muscles are also linked to other muscles via the spinal cord. These neural links or closed loops connect muscles working about a joint and they employ sensory and interconnecting as well as motor neurons. They are also monitored by connections back to the brain, termed gamma pathways, which determine the finer phases of physical movement; eg holding and using a light tool in a delicate process. The nature of the impulses themselves is not yet fully understood. Nearly all movement requires the use of a large number of muscles. For any movement the nervous system must receive, interpret and co-ordinate information from external sources and from our own bodies.

Movement which can be postural, manipulative or involve all the body parts in sequence, can be planned and executed as necessary, drawing heavily on neokinetic levels of learning which develop our habitual conditioned reflex responses. Neural interconnections to a muscle or between muscles are studied in kinesiology as reflex arcs and are the functional units of the nervous system that provide instant response from skeletal muscle.

A muscle is divided into two main kinds of tissue: a contractile portion between two tendons which are non-contractile. The two tendinous portions are generally designated the origin and insertion of the muscle. Muscles act as prime movers when they initiate an action: as antagonists when they act against prime movers to control the rate of action; as partial antagonists when they assist the prime mover to make one definite action; and finally as fixators or stabilisers when they hold parts of the skeletal frame firmly together whilst an action takes place elsewhere. Muscles rarely function alone, and most are composite in fabric and can undertake more than one function at a time. It is possible for one portion of a muscle to be active in a major role, whilst another portion is playing a minor role, and yet a third portion showing no activity at all. So that muscle action is often the co-operative action of a number of muscles, or some of their parts. Thus compositely the actions of separate muscles shade together to effect varieties of movement and to give them clarity and precision. Composite muscle action is often difficult to assess in terms of what number of muscles or parts of muscles are responsible for particular movements. As has already been shown, it is quite possible to use an alternative sequence of muscle actions to effect the same kind of movement, and can again be demonstrated in another manner by the achievement of subjects who have temporarily or permanently suffered partial locomotive damage. The composite actions of muscles suggest that when consideration is being given to the situation of a human operator, provision should be made for space to allow a shift of position so that fresh muscle relationships can be tried and the operator enabled to carry on. Shifts of position offset fatigue.

Antagonists pull against prime movers even in so-called rest. Some muscle energy is therefore always being expended. This quiet pull of muscle against muscle is a form of static contraction, sometimes referred to as 'tonus' and it adds to the general feeling of well-being in a healthy person.

Part of a muscle's duty in composite action is to exchange roles to prevent damage to a joint. Thus the flexor muscle of the arm, the bicep, becomes an antagonist to the tricep muscle a fraction of a second before full extension of the arm at the elbow is completed. The bicep pulls momentarily at the termination of the movement. This reciprocating pull is a conditioned

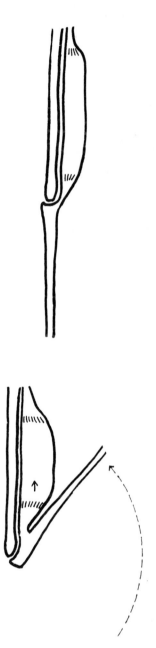

Figure 59
Diagram to illustrate the simple action of a muscle. When a muscle contracts it shortens its length. With its two ends attached to different bones by tendons, upon contraction one or the other of the bones must move by pivoting at a joint

reflex action in the functional unit serving these particular muscles. A further reflex action for the contraction of the antagonist muscle is initiated from the sensory endings in the muscle itself, the ligaments and other tissues binding the joint giving movement. In normal movement the reflex action that originates from joint tissue becomes the second means of deaccelerating the action of a joint.

Reduction of the use of joints in a limb may lead to economy of an operator's energy. For instance the use of the whole arm-reach without movement at the elbow or shoulder is much less tiring for short periods than the continuous use of all the intermediate joints; because the olecrannon process of the ulna fits into a corresponding fossa of the humerus when the elbow is extended, and the manner in which the two bones can rest together allows an amount of relaxation for the surrounding muscles. However in the consideration of the use of limbs by an operator a careful evaluation of their task is necessary, and the conservation of energy may be seen as secondary to the effectiveness of their operations.

Considerations towards the use of human limbs and joints in operator functions

Forces should only be moved by limbs whose joints are in a 'midway' or intermediate position (Figure 60). The 'midway' position for commencement of exertion allows the arcs to function in an optimum time period.

The resistance of the force may also be too great for the operator to overcome. Attempts to overcome resistance or force from the 'midway' position allow the operator to test their capacities naturally. Trial against resistance when the joint is near the end of its movement not only causes anthropathic effects, but positive damage.

There are further reasons why it is important that forces should be countered by limbs whose joints are in a midway position. Human bones, particularly the long bones in the arms and legs, together with muscles that power their movement and the joints that hold them together, can be considered to be acting as first or third class levers. (There is continuing debate, about the presence of second-class levers because they are extremely difficult to diagnose.) Be that as it may, the fact that we do use levers is relevant to the problem of deciding upon the most efficient means of using our physical powers. The power of a human lever in exertion depends on the power or the effort the muscles can apply to the bony lever to obtain equilibrium with, or to overcome, the resistance being applied to it, and the power that a muscle can offer is greater at the beginning phase of contraction than at the termination. Midway positions, where muscles are only partially contracted therefore give larger forces, or the possibility of more effort. (Figure 60.) The 'midway' position also allows prime mover muscles which give

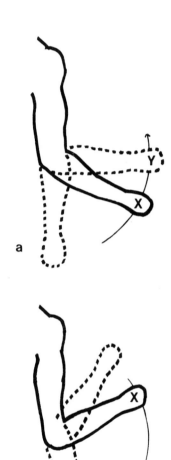

Figure 60
a illustrates the arm in a midway position towards full flexion. Power is greater at position **X** to overcome resistance than at **Y** where it would decrease rapidly

b illustrates the arm in a midway position towards full extension. Power is greater at position **X** to overcome resistance and it decreases to **Y** and beyond

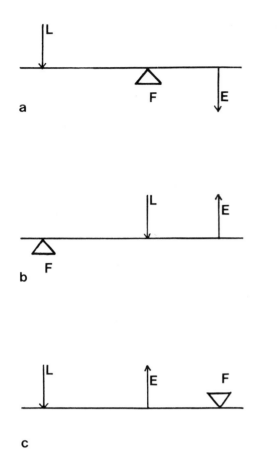

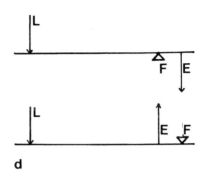

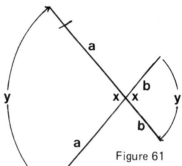

Figure 61

a, **b** and **c** show the three orders of levers:

L = The load or resistance to be moved

F = The fulcrum

E = The point of application of the effort or force

a is first-class, **b** second-class and **c** is third-class

d   illustrates how first-class and third-class levers can be effectively increased in length by the point of effort and the fulcrum being brought together

e   illustrates how the lengthening of a lever can increase the velocity at the load end

Levers **a** and **b** have the same angular velocity at **x**. **a** however is longer than **b**, and at the load end **y** **a** has a much greater linear velocity than **b**

An example calculated from human movement shows that the hand swung by the extended arm from the shoulder moves nearly three times faster than the hand swung by the forearm from the hip

power to a movement to operate fully; these muscles are called shunt muscles and are attached to the bone acting as a lever at some distance from the preferred joint towards the middle and end of joint movements. The moving bone is powered more by spurt muscles which are attached near to the preferred joint.

Third- and first-class lever lengths can easily be increased effectively; that is to say the points of the fulcrum and of the application of power can be kept together, whilst the application of the load or resistance can be moved further away. (Figure 61.) As the length of the lever increases so the velocity at the load end can be increased. (Figure 61.) This is precisely what has taken place in the evolutionary construction of man's limbs, and it is therefore possible for us to move optimum loads with speed.

With limb movement the correct speed for the action is important to the human operator. Speeds of operation which are too quick or too slow cause fatigue; the fatigue being generally caused by unnecessary complex muscle control due to unnatural movement and positioning. Complex muscle control of joint movements is wasteful and inefficient, when it only produces a small effect that could have been produced by simpler means. The operator being placed too near or too far

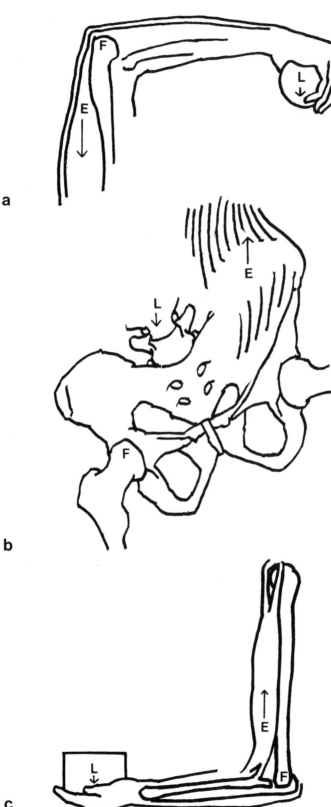

a

b

c

Figure 62 a, b, c
Anatomic examples of the three orders of levers

a The forearm working as a first class lever. Front view of the right arm holding a ball over the head

F The fulcrum is the distal end of the humerus bone which is being held upright

L The load is in the hand grasp plus arm weight

E The effort is produced by the contraction of the triceps muscle.

b The pelvis working as a second class lever. Front view of a person balancing on the right leg, the left leg being abducted.

F The fulcrum is the head of the femur bone

L The load is the weight of the upper parts of the body

E The effort is the production of phases of static and concentric contraction in the oblique abdominals and ilio-costalis muscles

c The forearm working as a third-class lever.

F The fulcrum is the distal end of the humerus bone

L The load is in the palm of the hand.

E The effort is the contraction of the biceps muscle which is over-coming the resistance of the load and causing movement upwards

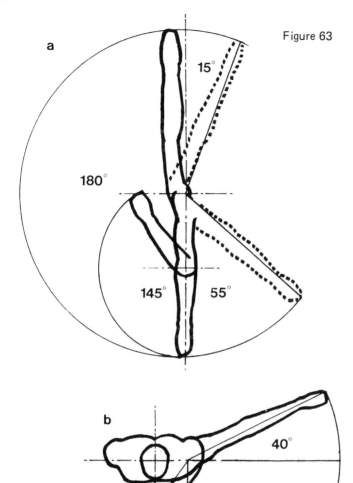

Figure 63

Figures 63, 64, 65, 66
Diagrams to show the capacity for angular movement of the limbs or limb segments. The angles stated have been estimated from a variety of surveys, and in most cases there is not just one angle of movement of which everyone would be capable. As with other forms of measurements there are gradations of values in terms of percentiles. For critical operations retesting would have to be carried out with the population of users concerned. The angles stated should cover the contingencies of everyday operations

a Shoulder flexion from the horizontal to behind the head. Shoulder extension to behind the centre line of the body. Elbow flexion

b Shoulder hyperextension behind the body, and shoulder adduction bringing the arm across the body

c Rotation of the elbow

d Elbow flexion towards the medial body line

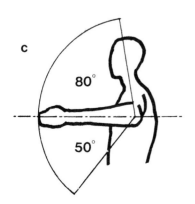

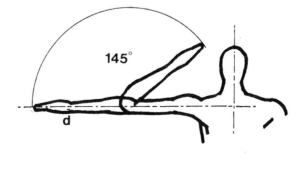

Figure 64

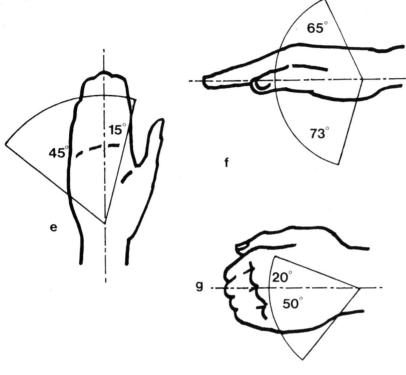

e Hand palm facing. Wrist adduction towards the body medial line, and wrist abduction

f Wrist dorsiflexion above the centre line, and palmar flexion. Forearm pronated

g Wrist flexion and extension with forearm supinated in the perpendicular plane

Figure 65

h Hip flexion above horizontal line, extension and hyperextension behind the line of gravity

i Hip rotation. Lateral and medial

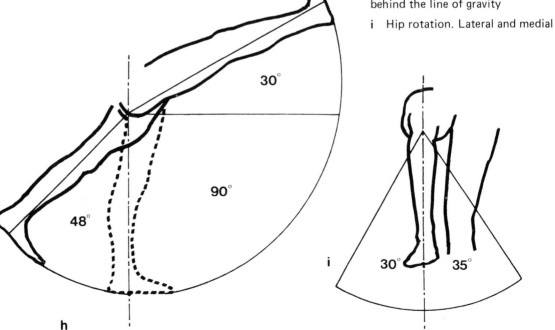

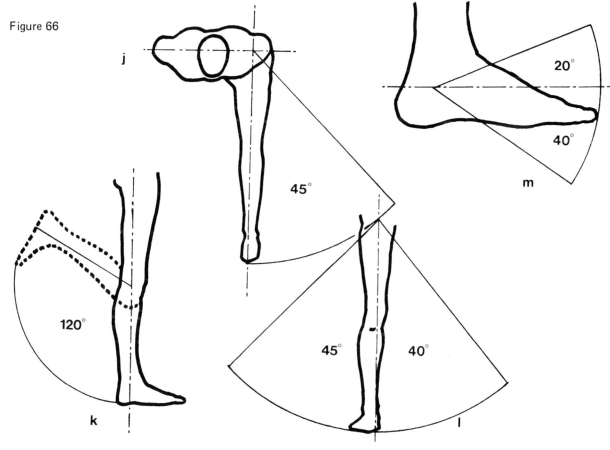

Figure 66

j   Hip swing. Lateral and medial

k   Knee flexion

l   Hip adduction across medial body line, and abduction

m   Ankle dorsal flexion above the centre line, and plantar flexion

away from controls are both positions where muscular and nervous energy would be wasted, for in both postures the limbs would be over extended or uncomfortably flexed, and their use in this position would make unnatural demands on an operator. For example, too great a flexion of the knee will restrict the smooth operation of ankle or foot movements. Too great a flexion of the elbow will restrict the smooth operation of wrist and hand movements. When the leg is partially extended foot movement on pedals are under far better control. When the elbow is partially extended, pronation and supination at the wrist is smoother, and the hand is able to offer a more deliberate form of control. For very delicate control movements the fore-arm may be retained near the operator's body and the hand held slightly below elbow level.

Body and limb dimensions related to enclosed and working environments

Movements of force, or movements with levers or wheels are less awkward and quicker if they can be arranged to move towards the medial line of the body or downwards following the force of gravity. It therefore follows that anti-clockwise movements are easy and efficient, particularly when arranged to take place in a vertical plane. (Figure 67.)

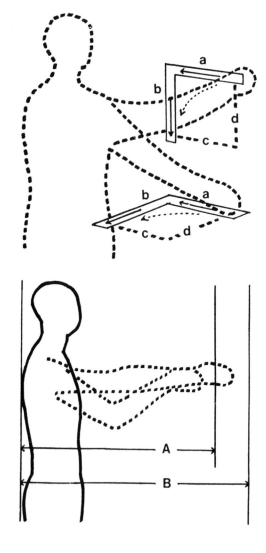

Figure 67
Control movements made inwards
and towards the body can be faster
and superior to movements made
outwards and away from the body
    They can be equally effective in
the perpendicular plane
    The control movements can
either be angular, or as shown by
the dotted arrows, follow a curved
path
    Paths **a** and **b** in both cases
would be actionable movements;
c and **d** return control paths

Figure 68
A favourable position for maximum
pull and push movements

A Distance towards which
maximum pushing could take place
is 737 mm (29 in.)

B Distance from which maximum
pulling could take place 864 mm
(34 in.)

Operational levers

Levers can be arranged to give a wide range of movement; the
greatest range of movement being obtained when they pass
across in front of the body, rather than towards or away from it.
Greater force can be applied by the second type of lever movement. Lever forces are not suitable for fine adjustment, but
they can give positive on/off switching or positive multiple
switching providing the range of intermediate contacts or
positions are neither too great in number or too close together.

   Large forces can be applied by a lever particularly when an
operator is standing, and greater again if the weight of the entire
body is allowed to contribute. For the reasons that have been
discussed elsewhere this is unwise and should not be allowed to
take place, and levers should be so sprung that only minimal or
medium effort is needed for their movement. A travel distance
of 102 mm (4 in.) can be quite satisfactory for a short lever.

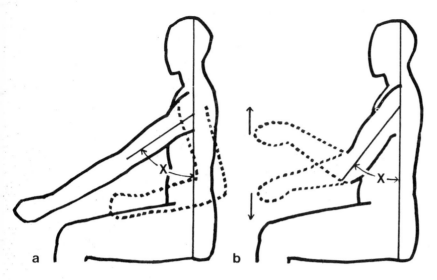

Figure 69

a Illustrates sitting position for operating a lever towards and away from the body. Difficult arm position is dotted. More favourable position for arm is when upper arm makes an angle of 60° (**X** on diagram) with centre line of trunk

b Illustrates sitting position for operating a lever up and down in the median plane of the body. Angle **X** (on diagram) of 30° of upper arm to the body is the most favourable position

Lever loadings for operator in sitting position. The lever moving up and down in the median line of the body

A good arm-position for this type of operation is when the upper arm is approximately 30° to the centre line through the trunk, loading amounts decrease above and below this angle. 5th percentile values for these two movements are:

Pull down 9.07 kg (20 lb); Push up 6.8 kg (15 lb). (Figure 69b.)

Lever loadings for operators. The lever moving towards and away from the body in the median plane with the lever in a more favourable position from the body

Values given for left arm. This is normally the weaker arm so that 5th percentile values provide maximums within the capacity of all for both arms. (Figure 69.)

Maximum pull only when arm is fully extended.
Maximum pull   5th percentile 22.7 kg (50 lb)
                    50th percentile 49.9 kg (110 lb)
Maximum push when arm is not fully extended.
Maximum push   5th percentile 18.14 kg (40 lb)
                    50th percentile 54.43 kg (120 lb).

Lever loadings for operator in sitting position. The lever moving towards and away from the body in the median plane

The maximum load for the most difficult position for the left arm for the 5th percentile operator is 9.07 kg (20 lb). More favourable positions for the left arm for the 5th percentile operator allow loadings of 13.6 kg (30 lb).

For push-pull levers 9.07 kg (20 lb) can be taken as a very safe loading; 13.6 kg (30 lb) therefore, can be taken as a safe loading figure and it is the one recommended.

Lever loadings for operator in sitting position. The lever moving across in front of the body laterally

5th percentile values for this movement are pulling in 6.8 kg

Figure 70
Lateral lever movements

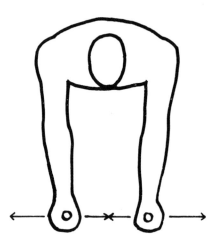

(15 lb); pushing out 6.8 kg (15 lb). (Figure 70.)

Pedals
Total leg strength has great power, and the force exerted by leg extension is greater than that obtained by flexion.

A large amount of force can be transmitted with leg extension by a seated operator using a back rest, although this cannot be sustained for any period of time. Although offering great power, leg and foot movements are not accurate. They are good for start/stop, on/off switching, steady pressure (although this may require footguards to maintain the foot on the pedal); and steady repetitive motion.

Pedals can either be operated by the whole leg and body (with or without back rest) or the foot pivoting at the ankle. (Figures 71, 72 and 73.) Foot pedals exert less force, but they do offer more accuracy because conscious control of the nerves and muscles acting on the feet can be co-ordinated to the degree of unison of action that will produce gradated pressures. But in

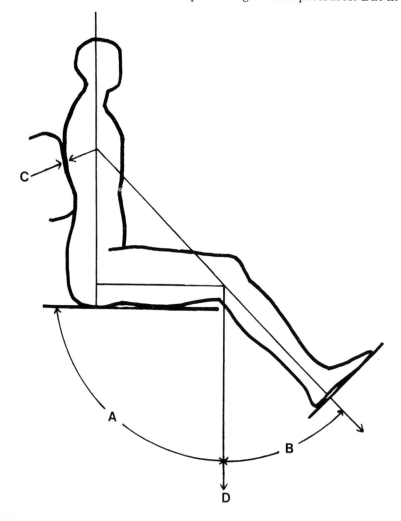

Figure 71
Leg position to achieve large forces of thrust
A = angle of 90°
B = angle of 45°

This position allows maximum thrust. Thrust at **D** would be less than 45.4 kg (100 lb). If any adjustment were made to the seat back and seat pan relationship, the 135° angle of the leg would still have to be maintained

C Counter pressure supplied by stable back rest

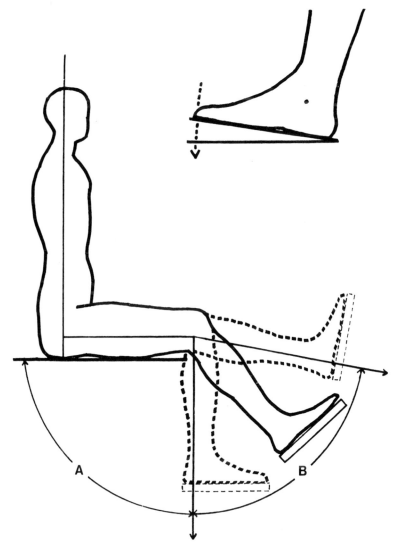

Figure 72
Foot pedals
Foot should make an angle of 90°
to 100° with lower leg. If the angle
of the pedal surface to the ground
is 15° or more a heel stop should be
provided

An optimum range of pressures
possible for continuous work would
be from 1.4 to 5.44 kg (3 to 12 lb).
A pressure of 9.07 kg (20 lb) or
more should be applied by the
whole leg

Figure 73
Leg positions for light thrusts of
13.6 kg (30 lb).
A = angle of 90°
B = angles between 90° and 170°

The pedals are at right angles to
the lower leg

Adjustable or sliding seating
should be provided for pedalling
positions

If possible pedals should be
spaced 305 mm/356 mm (12 in./
14 in.) apart, their centre being in
line with the median line of the
body

90°/100° should be maintained
between foot and operated surface
as with foot pedals

general the action of the legs and feet in the majority can in no
way approach the refinement of action and tactile awareness of
the arms and hands.

Pedals cannot be operated satisfactorily or comfortably from
a standing position. Both feet are normally used in standing to
preserve balance; if a floor pedal is used it should be at ground
level and have a very short travel distance, so that the operator's
two feet remain as near as possible to the horizontal.

Without undue effort any thigh and lower leg angle between
90° and 170° can offer a force of 13.6 kg (30 lb). (Figure 73.)

Much greater forces can be exerted by foot pedals, the effort
required being minimised by the operational leg being 'sighted'
in a particular angular relationship. Typically this is an angle of
135° between the thighs and the lower leg. (Figure 71.) With
this angle or slightly greater forces of 136 kg/181.43 kg (300/
400 lb) can be offered. Thrusts with the leg to achieve these

forces or greater are very quick and the counter pressure of a
back rest is needed. The angle of 135° is critical as a minimum;
there are degrees of tolerance towards a larger angle but the
limiting factor comes with most figure types before the leg has
reached two-thirds of full extension. This is because of the
lower leg's action as a third class lever in extension.

With effort larger forces can be exerted in this position, but
only for fractions of a minute.

Various angles of leg presentation and body stance can therefore present a gradation of forces from 13.6 kg (30 lb) to 136 kg
(300 lb) or more.

Percentile values for feet dimensions in millimetres and inches

	5th		50th		95th	
	mm	in.	mm	in.	mm	in.
Foot length male	248	9.8	267	10.5	287	11.3
Foot length female	223	8.8	241	9.5	259	10.2
Foot breadth male	91	3.6	99	3.9	109	4.3
Foot breadth female	83	3.3	91	3.6	99	3.9

Cranks A handle mounted on a turning wheel
Cranks up to about 102 mm (4 in.) in diameter give faster
speeds. After this size the speed decreases as the size of the
wheel increases. They offer a good degree of accuracy and
particularly the larger wheels can be used where the control of
adjustments can be arranged to follow continuously and full
circle. They can be used to move loads of 13.60 kg (30 lb) or
more. The shaft of the wheel can either face the operator or be
set parallel to the way he is facing. (Figure 75b.)

Handwheels
Used by both hands and give good operator stability. They offer
reasonable powers of adjustment through a turn of 90° or less.
They are best when placed facing an operator. When mounted
in a perpendicular position they can exert greater force when
turned in an anti-clockwise direction. They can be used to move
heavy loads in the same way that a lever can, and small wheels
in this position can be operated at a greater speed. However,
this could involve the use of the operator's body weight, and as
with lever operations, should be avoided if at all possible. They
are best used with moderate loads of 18.16 kg (40 lb) to 22.7 kg
(50 lb). Their diameter should be in the region of 305 mm
(12 in.) to 356 mm (14 in.).

Knobs
Knobs are used for a finger or precision grip for making fine
adjustments. Small sizes may be only lightly loaded. 38 mm/
51 mm (1½ in./2 in.) is a good size and offers smooth operation
and accurate adjustment. A 51 mm (2 in.) diameter knob can

Figure 74
Different finger and hand grips

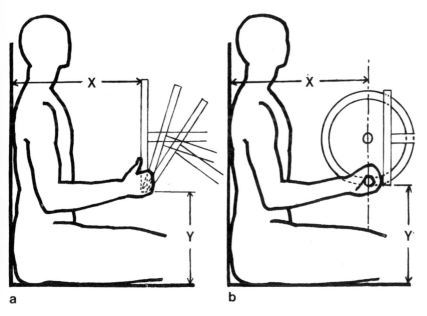

a

b

Figures 75a and b
a Suggested clearances for hand-wheels: **X** = 381 mm (15 in.) **Y** = 229 mm/254 mm (9 in./10 in.)

b Suggested possible positioning for a small crank wheel **X** and **Y** are still 381 mm (15 in.) and 229 mm/254 mm (9 in./10 in.) respectively. Two positions for crank wheels are shown. Special heavy duty clothing might call for revision of these clearances

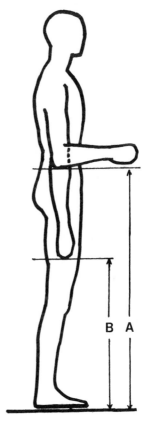

have loadings from 0.9 kg to 3.63 kg (2 to 8 lb). (Figure 74d.) A 76 mm (3 in.) diameter knob or more requires operation by wrist movement. (Figure 74e.) Knobs of this size may be more heavily loaded, and between 76 mm (3 in.) and 127 mm (5 in.) can take 6.8 kg/9.07 kg (15 to 20 lb) loading. Knobs greater than 127 mm (5 in.) become impossible to hold easily to produce a torque by operators with 5th percentile hand widths. Greater torque values must be produced by handwheels starting at 254 mm/305 mm (10 in./12 in.) diameters.

Knobs must be deep enough to be gripped well. Generally this means a depth of slightly less than 25 mm (1 in.): 25 mm (1 in.) being the mean length of a first phalanges section of a finger. (Figures 74a, b, c, d, and e.)

Hand or power grip 44 mm (1¾ in.) diameter. (Figure 74f.) Light finger control 10 mm (³⁄₈ in.) to 16 mm (⁵⁄₈ in.) diameter. 0.45 kg/0.68 kg (1/1½ lb) loading. (Figure 74c.) Single push button control 0.68 kg/0.9 kg (1½/2 lb) loading. (Figure 74b.) Multiple switches. Power hand grip 44 mm (1¾ in.) diameter. (Figure 74f.) Toggle switches. 25 mm/51 mm (1 in./2 in.) high 0.23 kg (½ lb) loading. (Figure 74a.)

Figure 76 **Switches and handles**
General-purpose switches and door handles are convenient if placed somewhere between elbow height **A** and finger tip height **B**. Taking the two smallest values for these distances we can find we have for women **A** 559 mm (34 in.) and **B** 610 mm (24 in.). This gives a band of 254 mm (10 in.), which is 610 mm above floor level, in which switches and handles can be placed, conveniently for the tallest person

Percentile values for hand dimensions in millimetres and inches

	5th		50th		95th	
	mm	in.	mm	in.	mm	in.
Hand length male	178	7	191	7.5	203	8
Hand length female	165	6.5	178	7	191	7.5
Hand breadth male	94	3.7	104	4.1	137	5
Hand breadth female	76	3	89	3.5	102	4

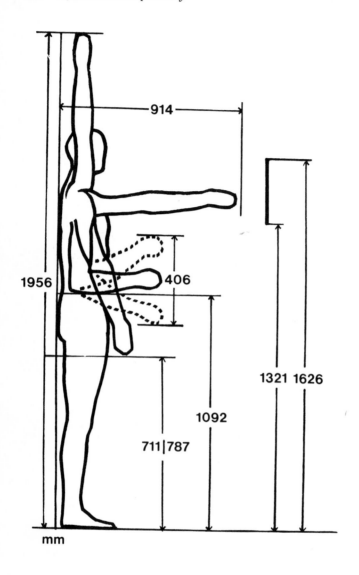

mm

Figure 77
Reach and stretch dimensions discussion
1956 mm (77 in.) is the maximum stretch reach above the head for the 5th percentile, it would also allow 25 mm (1 in.) or slightly more clearance for standing height for a 95th percentile man. 914 mm (36 in.) is the minimum distance that would allow all arm lengths to move unimpeded at full stretch. 1321 mm/1626 mm (52 in./64 in.) is the position for controls and small dials that could be seen clearly by most men and women. 1092 mm (43 in.) would allow clearance for most elbow heights. Hand height is difficult, but 406 mm/787 mm (28 in./31 in.) would be convenient for many people; something like 22.7 kg/ 27.2 kg (50 lb/60 lb) pressure can be applied by the hand from this position. The 406 mm (16 in.) range of hand position in the perpendicular plane is the maximum in which the hand can be manipulated without moving the elbow forward

Force exerted by squeezing the hand or handgrip
    Male   5th percentile 25 kg (55 lb)
    Female   5th percentile 22.7 kg (50 lb)
So that a safe maximum for a spring loaded grip or brake might be something in the region of 13.6 kg (30 lb).

Forces of 40.8 kg (90 lb) to 49.9 kg (110 lb) can be exerted by squeezing the hand, but these forces are approaching 95th percentile values.

Joysticks

Joystick operation involves movement of the arms and hands near the body and thighs. The loading for this type of control for 5th percentile operators is 6.8 kg (15 lb). The advantage of the joystick is its range of movement in an area in front of the body: the exertable force taking second place. Taking these

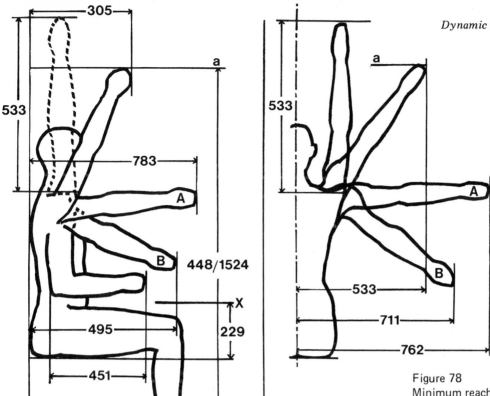

Figure 78
Minimum reach distances for hands clasping controls based on 5th percentile figures. **A** is reach at shoulder height. **B** is reach at waist height. **X** is the underside of a work surface and 229 mm (9 in.) is given as the minimum clearance for the thigh. 451 mm (16 in.) is given as the maximum clearance for the forearm. In general the 5th percentile reach distances, which are nearly the shortest reaches, should always be used. All critical controls should be placed opposite the operating limb. Forward reaches can be improved by up to 152 mm (6 in.) by bending the trunk forward. This however decreases reach distances above the head. Seat height has not been stated in this diagram, but it would be normally estimated as varying from 381 mm (15 in.) to 457 mm (18 in.). The seat-height/work-height relationship is much more important than seat-height/floor-relationship. Adjustable seating and adjustable foot rests can be used to maintain optimum seat height/work height

a on both diagrams is 1448 mm/ 1524 mm (57 in./60 in.)

facts into account 6.8 kg would be a good joystick loading. Lateral and fore and aft travel should be about 152 mm/178 mm (6 in./7 in.) in each direction. These distances relate well to 5th percentile arm reaches and the possible positioning of pedal controls. Maximum trunk depth, however, must be considered when positioning a joystick. Therefore adjustable seating would offer the best solution for fitting an operator to this type of control.

Seats—Seating: General
Sitting is a means of changing posture and bringing rest. During the course of a day we continually change posture to reduce fatigue. Probably, of all the possible postures, the most effortless is the fully reclining semi-foetal position often adopted in sleep.

Sitting can be a tiring and painful business on a poorly designed seat. A good seat should allow for movement or a change in the sitting posture; there must be space for easement to maintain the best sitting posture for a lengthy period, but there should be enough control from the seating surfaces to effect the relief of body weight and give a sense of stability. A seat needs to give rigid support but not rigid confinement. It should in particular support the thorax and pelvis and help to maintain the angle of the spine between. Therefore, the design of back-rests is important. A low seat is better than a too high one, but no seating should be so low that legs are stretched far out in front of it without leg or foot support. (Figure 79b.) A working chair of correct height will allow leg movement backwards with the feet flat to the ground. A too-high chair will cause unnecessary pressure under the front of the thighs, and give pain and discomfort, for the soft tissues of the thigh is not meant to be a weight bearing area, and its compression can only lead to traumatic effects. Sprung upholstery of a cushioned surface can be good, but it is not a substitute for incorrect chair height.

Quite involved adjustable seating is used in industry and health-care work, and it is always necessary where seating dimensions are critical. Adjustable head, back, arm and foot rests can all lend support to a more effective use of the body when seated. Adjustable seats and supports can supply a controlled mobility.

Collections of measurements may be the beginning of the design process for seating, but only practical trials of prototypes will suggest the necessary modifications that make a successful design. It is useful to take past seating solutions that have reference to current problems and view them together, superimposed over one another by photographic methods. Critical measurements in the current problem can be tried against them. Photography can also supply a rapid record of the varying postures a trial population naturally adopts when using the prototype. It will be found that the study of the variation of posture by trial users will suggest modifications to features and dimensions.

There is a wide range of types possible in seating design, and with the absence of a specific demand the possible use of a new design should be carefully considered.

Seating: height

Of all the dimensions of seating, seat height is one of the most important. A chair height must not be so high that tissue in the distal and posterior region of the thigh is compressed and the seat front edge forced of act in the manner of a tourniquet to the blood supply of the legs. (Figure 79a.) Every seat height should be able to fit the shortest length of any lower leg. An acceptable height fit should allow a small space under the distal

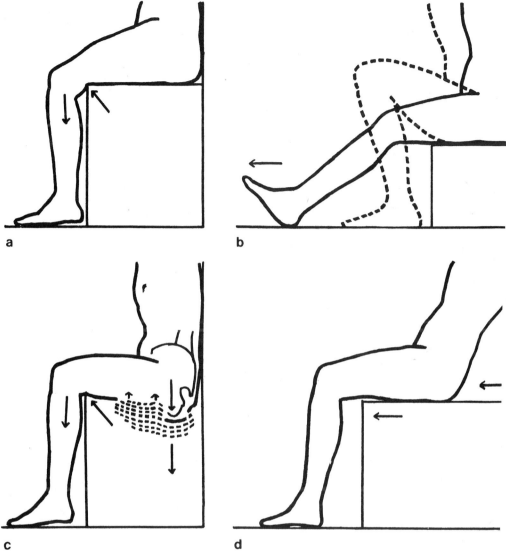

Figure 79

a Seat too high. Front edge of seat surface cutting into back of lower thigh

b Seat too low. Legs slide forward unsupported without stability at the foot. The dotted line shows low seat crouch

c Upholstery too soft. Weight not taken by ischial tuberosities. Front edge of seat surface cutting into back of lower thigh

d Seat surface too long. Seat edge cuts into back of knee and sitter slides forward as a consequence, the back rest then being unused

portion of the thigh as a hanging space for soft tissue, when a seated person would have their knees at right angles and feet flat on the ground. (Figure 80.) By reason of the effect of tourniquet a tall person can find more comfort in a low seat than a short person can from a high seat. (Figure 79a and b.) In general it a seat is too high a person's body will slide forward and any special planning for the depth of a seat or its backrest will be lost. If a seat is too low the seated posture will assume a forward crouch with most seating benefits lost. Adjustable seat height should offer a range of from 368 mm (14½ in.) to 470 mm (18½ in.) for men, and from 356 mm (14 in.) to 445 mm (17½ in.) for women. (Figure 81.)

Seats are often found to be 457 mm (18 in.) high, but this is too high for the majority, and too high for a seat for a working

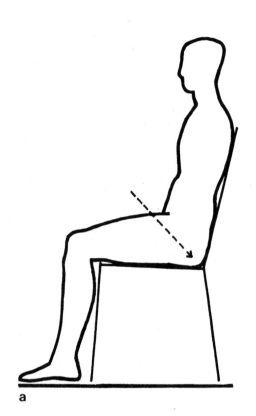

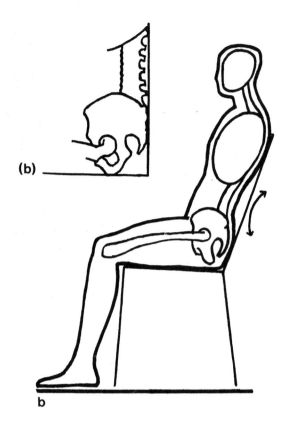

position for long periods. 432 mm (17 in.) has been recommended, but this is still a height which is excessive for women.

A reasonable fixed seat height for men would be 419 mm (16½ in.) but for a woman 387 mm (15¼ in.). 406 mm (16 in.) offers a good compromise and is, in fact, a good height for adults, and the optimum sitting height.

For general leisure purposes seat height can be reduced still further to something under 406 mm (16 in.) with a minimum at 381 mm (15 in.). Where the lower leg is angled forward and the seat is tilted, as in most driving positions, the height of the seat can be 356 mm (14 in.) or less. But for quality driving a fully adjustable seat which offers each individual the best compromise between reachability of controls and maximum vision is best.

Seating: foot-rests

Foot rests must be used where the seating positon carries the whole body above normal sitting height level. If they are used for this purpose or any other they should always allow the angle between the lower leg and the base of the foot to be normal or about 90° to 100°. (Figure 82.) Either when used as a built-in fixture on a piece of apparatus or a chair, or separately as a foot-stool, if the surface is greater than 15° from the horizontal a heel stop should be used. (Figure 82b.) The overall surface needs to be large enough to rest the whole foot.

Figure 80
a A slightly tilted seat base and angled back rest to help comfortable 'settling' in a chair

b Angled back rest assisting the lumbar curve to maintain its natural curve or 'arch'

Insert (**b**) shows the straightening of the lumbar spine when there is an angle of 90° or less between the seat surface and the back rest

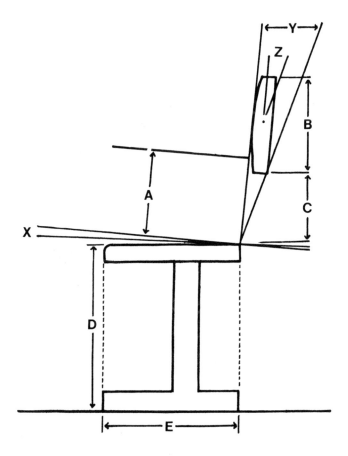

Figure 81
The adjustable chair
The great value of the adjustable
chair is that it can be altered to
allow for the very small differences
required in a chair from person to
person. For important and high-
level quality operating sitting
positions, it cannot be bettered.
An adjustable chair may be able to
take care of individual seating
troubles that an anthropometric
study may fail to do, and that our
sitting habits require. The chair's
mechanism should allow it to adjust
easily and quickly.

A Arm rest heights 203 mm (8 in.)
to 216 mm (8½ in.). If it could be
adjustable a range of from 191 mm
(7½ in.) to 254 mm (10 in.) is
recommended.

B Back rest height 102 mm (4 in.)
to 152 mm (6 in.) for men. 203 mm
(8 in.) for women. It should be able
to tilt independently **Z** from
angular back rest movement **Y**.
Its surface to be slightly convex

C Back rest height from seat
surface 152 mm (6 in.) to 191 mm
(7½ in.) for men. 127 mm (5 in.) to
191 mm (7½ in.) for women.
Height adjustable if possible

D Seat surface height 356 mm
(14 in.) to 483 mm (19 in.). A wide
range which 'fits' most lower leg
sizes

E Seat length or depth. 343 mm
(13½ in.) to 381 mm (15 in.). This
presents a dimension which should
be comfortable for a wide range of
men and women. It should not cut
the back of the knee. It could be
extended to 406 mm (16 in.).
Reduced to 305 mm (12 in.) it
could give an operator greater
mobility, but is then something of a
momentary perch and would
require foot rests

X Angle of seat surface. From 3°
to 5° or from 0° to 5° degrees

Y Angle of back rest. From 95° to
115° from seat surface

Seating: arm rests

Arm rests are often used as the bases for arm leverage in working positions, or for getting in and out of a chair. The critical
distances are those between the arm rests, and between the seat
and the arm rests. Arms should be able immediately to find the
rests without sideways search, and the arm rest height should be
able to accommodate upper arm height comfortably without
levering the trunk out of the chair. (Figure 83.) Between
extreme figure types there is no ideal height for arm rests. In
critical situations an adjustable rest would have to be designed.
The arm rest can be very useful where an operator has to use
fine and sensitive finger and wrist operations unhampered by
arm weight. Like seat surfaces they should not be too soft or
too smooth.

A useful distance between arm rests is 483 mm (19 in.). A
useful workable height is 203 mm (8 in.), but this might have to
be adjusted to 216 mm (8½ in.) if padding rests were used. A
minimum and maximum range for arm rest heights is 190 mm
(7½ in.) to 254 mm (10 in.).

Seating: width

This dimension only matters when a minimum allowance has to
be thought of for hip width and the spread of the buttocks, or
for a shifting of position. Having looked at these factors a

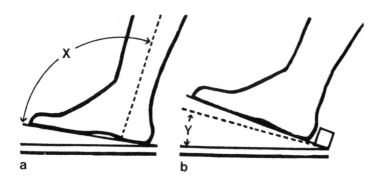

a   b

Figure 82
a X angle of foot rest surface to lower leg. 90° to 100°

b If the angle of a foot rest to the ground is 15° or more at Y, then a heel stop should be provided

comfortable width is 457 mm (18 in.), and nothing less than 432 mm (17 in.) is tolerable for lengthy periods. In special circumstances space would have to be allowed for clothing, particularly where seats are confined by arm rests. (Figure 84c.)

Seating: composition (hardness or softness) and shape
Common experience and testing has shown people seated in softer seats give a better performance; that is, greater efficiency, for a longer time. Anatomically it is the ischial tuberosities that carry the body's weight in the seated position. (Figure 80.) The tissue region over these bony protuberances has a superior blood supply to other tissue in the buttock formation. There-fore, because of the foregoing reasons, although seat surfaces need to be soft, they must at the same time be able to exert counter-pressure and not submit weakly to the body's weight. Fabric covers should be porous and allow moisture to pass through. A depression of 13 mm (½ in.) in a padded seat is enough. More than that does not allow the average ischial tuberosity to take the majority of the weight and the surround-ing tissues become too compressed. (Figure 79c.)

Figure 83
If arm rests are too high their purpose is defeated for they tend to lever the sitter out of the chair, or bring unnatural pressure on the shoulder joints

For short-period sitting a seat surface can be completely firm, providing space is left for shifting the body position. Allowance for adjustment of position delays fatigue in sitting.

Sculptured seat surfaces with slight depressions carved to take the bony protuberances have been tried. Seat designs based on this idea can only expect limited use.

Seat surfaces should not be too highly polished. A textured or rough surface is best.

A rounded seat front edge is also good.

Seating: depth
The depth for a seat does not present any great problem. The minimum back of knee to buttock measurement on a woman is about 406 mm (16 in.); the maximum of the same measure-ment on a man is about 508 mm (20 in.) or more. Taking the smallest measurement as a guide 381 mm (15 in.) for seat depth would be suitable, and this is a measurement which in fact is often used. (Figure 79d.)

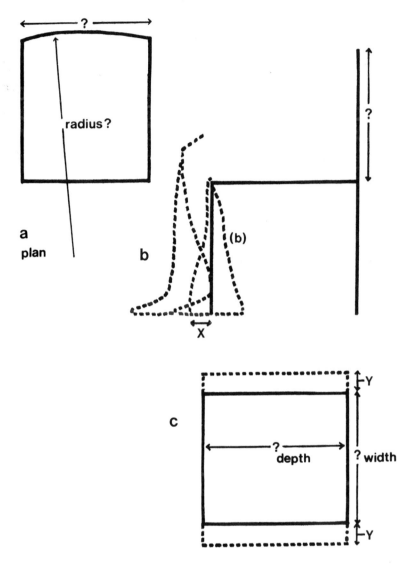

Figure 84
The consideration of space in seat design

a Seat back width at shoulder level or above for leisure and lounge seating 508 mm (20 in.) or more. 508 mm (20 in.) incorporates shoulder widths of the 95th percentile for men. 533 mm (21 in.) is the 99th percentile value. Curvature slight for leisure. Radius of curve two and half seat lengths. Seat back width at 356 mm (14 in.) high 381 mm (15 in.) to 466 mm (16 in.)

b Seat back height. 356 mm (14 in.) high for utility, 508 mm (20 in.) to 635 mm (25 in.) for shoulder support.
(95th percentile value for men's shoulder height is 635 mm (25 in.). 559 mm (22 in.) would be maximum for driving seat where all round vision is required. 889 mm (35 in.) for back and head support

(b) Allow for movement of legs under chair for alert and working seating. Also allow for 102 mm (4 in.) between front of chair and heel **X**

c Seat width. 457 mm (18 in.) for comfort and where there are enclosed arm rests it is a minimum. 483 mm (19 in.) better. Arm rest widths 102 mm (4 in.) or more **Y** Seat depth 381 mm (15 in.) is safe. 508 mm (20 in.) for lounge or leisure, but only when front of seat height is reduced to about 356 mm (14 in.)

Chairs made from hard materials, or with firm padded surfaces, are better with depths of 381 mm (15 in.) or less. In every case they would be improved if the front edge of the seat were rounded. With easy chairs depth can be generous as long as the upholstery is soft and the seat height is low.

Seating: back-rest
Back-rests may be constructed at a varying series of angles from a seat surface, depending on whether an upright seated position or a reclining or semi-reclining position is required. (Figure 85.) A slightly tilted back-rest helps one to settle comfortably in a chair and prevents a gradual slide forward of the body. (Figure 80a.) Without the tilt of a back-rest the lumbar curve is unnaturally flattened and strain is placed on the intervertebral lumbar discs and ligaments. Back-rests should give full back

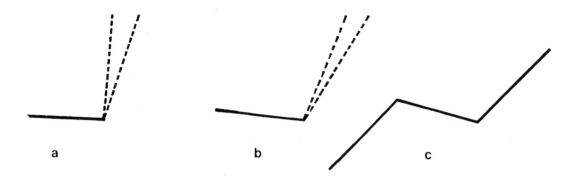

a   b   c

support, and if possible not interfere with arm movements. (Figure 80b.)

An angled back-rest also assists the force of gravity to settle the user's body into the chair and keep in such a position that the lumbar support section of the back-rest is used to the full. There are divided opinions on the use of a tilted seat base to further aid the cause. Probably in an upright working position adopted in offices a sloped seat would only be a disadvantage. A tilted seat base has an advantage in a semi-reclining position where the angle of the back-rest tends to push the trunk forwards. The possible angles for a back-rest from the perpendicular in the sagittal plane, for the working upright position, varies from 5° to 20°. (Figure 85a.) For comfortable seating this angle could be increased to 35°, but this would then be combined with some angling of the seat base from some 5° to 7°. (Figure 85b.)

The upright seated position is one that can be maintained for only a short time without support, and a back-rest can aid stability of the trunk and prolong the inevitable arrival of 'sitting-fatigue'. A good back-rest should allow the back freedom to be 'arched' occasionally and improve the lumbar

Figure 85
Back rest angles

a   Alert or attention sitting

b   Every day, conference seating, relaxed travelling position. Back rest to horizontal level 110° to 120°. Seat surface angle 5°/7°

c   Reclining in comfort. Back rest to horizontal 135°

Angles between seat and back rest, and seat and leg rest 120°

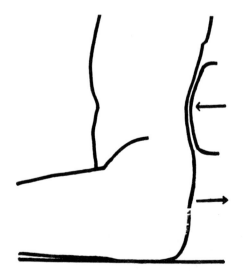

Figure 86
Back rest for a working position. Upper limit should be below shoulder blades so that movement of the shoulders and the arms is permitted. Thus the back rest becomes a lumbar spine support. About 305 mm (12 in.) to 356 mm (14 in.) wide gives free space for elbows

The gluteal region should have space to move and press backwards under the back rest

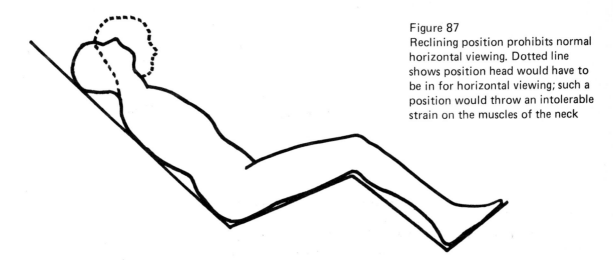

Figure 87
Reclining position prohibits normal
horizontal viewing. Dotted line
shows position head would have to
be in for horizontal viewing; such a
position would throw an intolerable
strain on the muscles of the neck

curve. (Figure 80b.) Its shape should be designed in such a fashion that is supports the lumbar region of the back, and if it should be of sufficient height, allowance made for shoulder blade support. The support for the lumbar region is of prime importance. It should be raised clear of the sacral region, or else its effect will be lost, 152 mm/178 mm (6 in./7 in.) being a good height for its lower border. (Figure 86.) In an easy chair specially constructed for a woman this could be raised to 191 mm (7½ in.). The height of the lumbar rest itself might be from 102 mm to 152 mm (4 in. to 6 in.); raised up to an 203 mm (8 in.) height if specifically for a woman.

Space for the free movement of the legs under the seat also assists the maintenance of the lumbar curve during sitting, because backward movement of the legs relaxes the posterior muscles of the thigh and allows the pelvis and sacral spine to rotate and maintain a normal relationship with the lumbar spine.

A fully comfortable and ideal back-rest might be one where the surface is moulded to accommodate the spinal profile to some extent. Such designs have been advanced from time to time with some success. The main advantage proffered by the idea of a continuously moulded back-rest, is that of being able to change the sitting posture from time to time whilst retaining back support. The shoulder support area in this type of design should make a minimum angle from the perpendicular of 15°. For complete relaxation the breadth of such a back-rest should be 508 mm (20 in.).

Reclining
The reclining or semi-supine position is a favourite for relaxing the whole body. Keegan's angle of 130° between the seat and back-rest is the position which rests the muscles most. As the back-rest seat angle is increased, so there must be a corresponding increase in thigh lower leg angle. This is not recommended if vision and full head movement are required. (Figure 87.) It

prohibits watching television or normal level view. Vision can be improved by increasing the angle at the knees, but moving the head can be tiring as its rotation is decreased by the amount it has to be pulled forward and into an upright position.

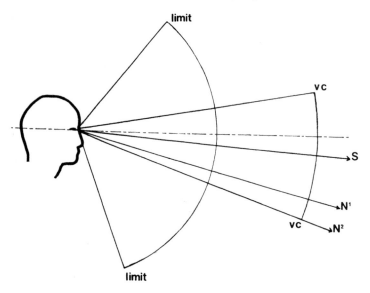

Viewing

If the head is not restricted the field of vision is large because slight movements of the head can appreciably widen the field of view. The conformable nature of the lens of the human eye always keeps the image in focus. The range of this agreement of the lens with the image allows us a comfortable minimum viewing distance of 406 mm (16 in.) (the optimum is about 533 mm (21 in.) up to a distance of some 6096 mm (20 ft.) or optical infinity).

The accompanying illustrations demonstrate the cones of human vision and viewing angles. (Figures 88 and 89.)

We gain a great deal of information by the use of our eyes, which can with training detect the very slightest tonal difference. We are able to identify a wide range of textures and solid forms, though an advanced visual skill must be developed by training. Apart from the head movements (Figures 89 and 91) given, and the various limits of vision, the eyes themselves are capable of movement. The rotation of the eyes adds to our ability to scan objects and the subtleties of their reflective surfaces.

Figure 88

Diagram to illustrate lines of sight and elevation of visual field

S is the standard line of sight and is 5° below the horizontal

N¹ is the normal line of sight when standing with the gaze undirected 15° below horizontal

N² Shows a further drop of the line of sight when sitting with the gaze undirected. 20° below the horizontal

We are a glancing-down animal by the construction of our eyes, which point below the horizontal. So that any horizontal scanning or looking up requires flexion of the neck and the use of muscles. **vc** is the visual cone 15° each side of the standard line of sight

Limit. Marks an area 50° above the horizontal and 70° below which can be viewed by eye movement alone

Control panel faces should be at 90° to the directed line of sight. So three possible control panel angles are: at S 5° to perpendicular, at N¹ 15° to perpendicular, and at N² 20° to perpendicular

5th and 95 percentile values for the heights of line of sight for standing and sitting positions in millimetres and inches

	5th MALE 95th		5th FEMALE 95th	
STANDING	1506 mm	1714 mm	1402 mm	1670 mm
	59.3 in.	67.5 in.	55.2 in.	63.4 in.

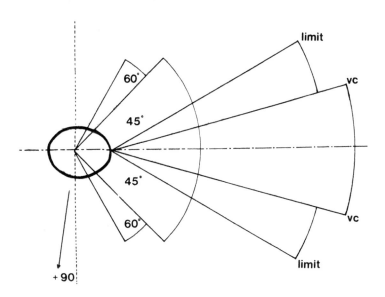

	5th MALE 95th		5th FEMALE 95th	
SITTING	745 mm	856 mm	663 mm	780 mm
From seat	29.3 in.	33.7 in.	26.1 in.	30.7 in.
surface				

Figure 89
Diagram to show plan of visual field and degrees of neck rotation

45° is the degree of natural head movement

60° is the degree of head movement achieved by conscious force

vc visual cone 15° each side of medial plane, is the area of vision which can be scanned by normal eye movement

Limit marks an area 30° each side of the medial plane, which can still be scanned by eye movement alone

The maximum viewing angle can therefore be something like a sweep of 180° when eye and head movement are used together. That is 60° + 30° either side of the medial plane

+ 90 indicates a 'out of the corner of the eye vision', achieved by eye movement alone

Figure 90
Diagram to indicate distances for the positioning of various types of information from the eyes

a 406 mm (16 in.). Minimum reading distance for print and instructions.

b 508 mm (20 in.). Good reading distance for general instructions.

c, d 533 mm/610 mm (21 in/24 in.). Distance for dial readings. Dials should be kept very simple, with the minimum of markings. Good face diameters would be 64 mm/76 mm (2½ in./3 in.).

e 737 mm (29 in.) is the furthest distance for the display of information

Working plane surface heights

Heights for working plane surfaces for industrial use are impossible to define and place in a general table. Machinery heights vary greatly but it is advantageous to mount it lower rather than too high. Elbow height provides a measurement of the upper limit for working comfort. Heights lower than elbow height allow more force to be used. Individual types of manufacturing tasks will need greater or lesser amounts of space and height because of the varying range of body movements they require. Standing positions offer greater scope than seated ones. A greater range of heights can be used, along with larger controls and greater force, the only limitation then being the distance of instructions or gauges from the eye.

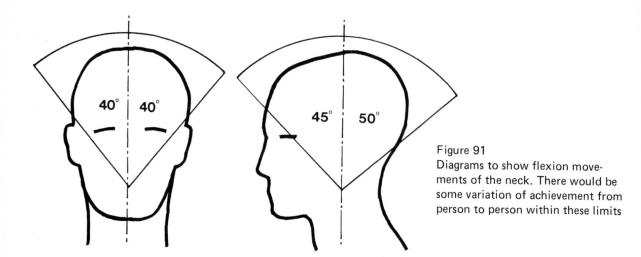

Figure 91
Diagrams to show flexion movements of the neck. There would be
some variation of achievement from
person to person within these limits

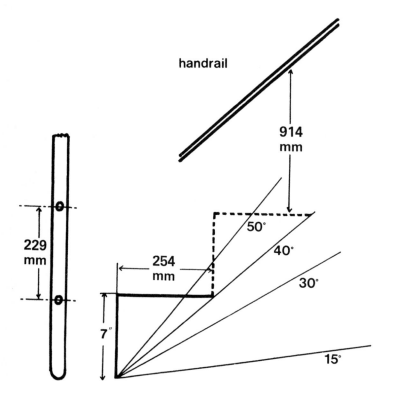

handrail

Figure 92
Stairs and ladders
Stairs are comfortable set at above
40°. Beyond 50° they are really
becoming ladders. Under 30° they
might be replaced by a ramp or
moving slope

Stair rises can be about 178 mm
(7 in.), the variations being between
152 mm (6 in.) and 203 mm (8 in.);
229 mm (9 in.) is too great. Stair
treads can vary from 241 mm
(9½ in.) to 305 mm (12 in.); they
may be greater than this

Stair widths can start at 838 mm
(33 in.). A 1219 mm (48 in.) width or
more is required when two people
might be passing frequently

Handrails should be between
864 mm (34 in.) and 940 mm
(37 in.) high

Ladders rungs should be spaced
229 mm (9 in.) or so apart.
203 mm (8 in.) or less is too
small and anything more than
305 mm (12 in.) can be tiring

Openings to lofts need to be about
510 mm (2 ft) square, the
dimension which accommodates
the shoulder width may be
reduced to 508 mm (20 in.) if
necessary

Height limits for various functional planes
1880/1981 mm (74 in./78 in.) Door and window heights.

1626 mm (64 in.)
1524 mm (60 in.) Comfortable maximum height to reach to.

1372 mm (54 in.)
1321 mm (52 in.) Height over which most can see. Height for central displays or internal view over partitioning.

1118 mm (44 in.)
1041 mm (41 in.) Highest counters.
965 mm (38 in.) Work benches: a good working height.
Above the mean value for elbow heights for standing women.
914 mm (36 in.) Popular work bench and sink heights. The counter height for women, and the height for standing ironing boards.

864 mm (34 in.)
838/882 mm (33 in./32½ in.) Food preparation surface heights.
762 mm (30 in.) Craft table tops and work room surfaces.
762/711 mm (30 in./38 in.) Average hand height standing.
737 mm (29 in.) Maximum for desk tops. Can sit and lean.
737 mm or 749 mm (29 in. or 29½ in.) for dining room tables.

Maximum height for tops of controls or bottoms of wheels for seated operator.
711 mm (28 in.) General table height.
686 mm (27 in.) Minimum working sitting height for general tasks, desk or control desks. Sitting at ironing boards.
660 mm (26 in.) Low working height. Typing tables.
635 mm (25 in.)
648 mm (25½ in.) Accommodates all knee heights including value for 95th percentile for men.
610 mm (24 in.) Minimum height for tops of controls or bottoms of wheels for seated operator.

Figure 93
Clearances for seated operators
Minimum sitting clearances plan

A 305 mm (12 in.)

B 406 mm (16 in.)

C 1168 mm (46 in.)

D 1270 mm (50 in.)
Person passing

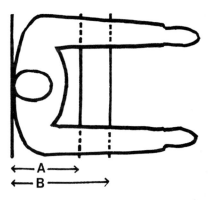

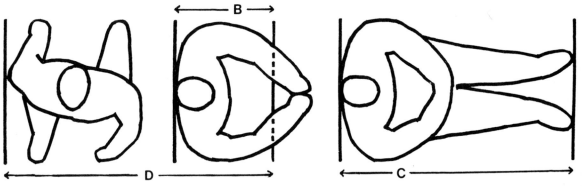

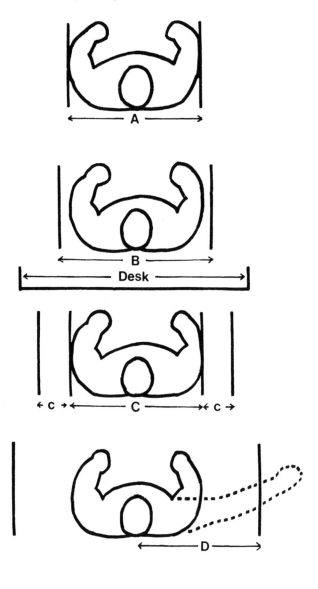

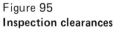

Figure 94
Minimum sitting clearances plan

A 610 mm (24 in.)

B 660 mm (26 in.)
Eating 711 mm (28 in.)
Theatre 762 mm (30 in.)

C 610 mm (24 in.)
c 152 mm (6 in.)
Desk 1067 mm (42 in.)

D Side counter 508 mm (20 in.)

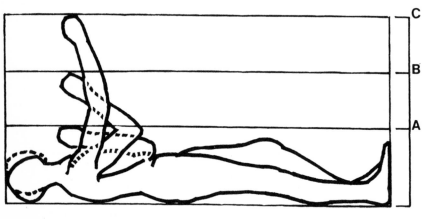

Figure 95
Inspection clearances
Lying positions for inspection and repair. Minimum heights are given; the length is 1930 mm (76 in.)

A Lying for inspection
Height 457 mm (18 in.)

B Restricted space for small tools and minor adjustments. Power from elbow extensions not possible. Height 610 mm (24 in.)

C Space for reasonable arm extension 152 mm/203 mm (6 in./8 in.) length; power tools could be used. Height 813 mm (32 in.)

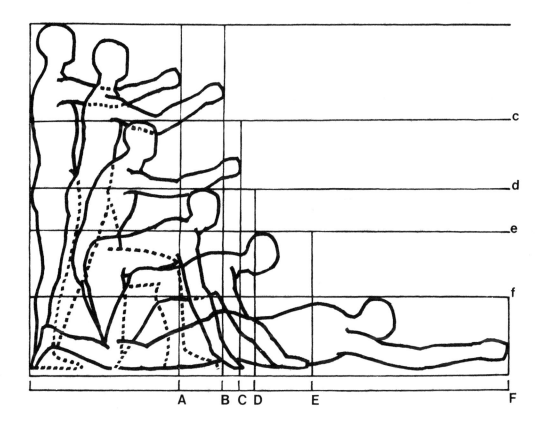

Figure 96
Minimum space allowances for the human body in a variety of postures

A Standing. Space for reasonable arm extension. Depth 762 mm (30 in.)

B Standing. Legs braced apart. Small power tools could be used. Depth 1016 mm (40 in.)

C Kneeling. Depth 1168 mm (46 in.)
c Height 1219 mm (48 in.)

D Crawling on all fours. Depth 1219 mm (48 in.).
d Height 889 mm (35 in.)

E Flattened crawling. Depth 1473 mm (58 in.).
e Height 813 mm (32 in.)

F Prone arm reaching forwards. Depth 2438 mm (96 in.)
f Height 457 mm (18 in.)

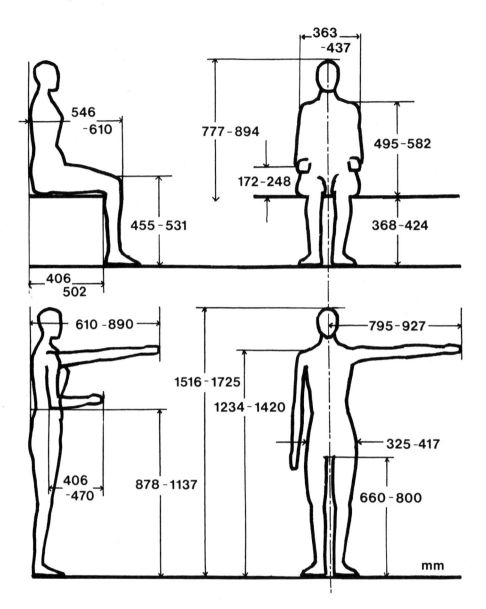

Figure 97
Dimensions of female body given
in 5th and 95th percentile values

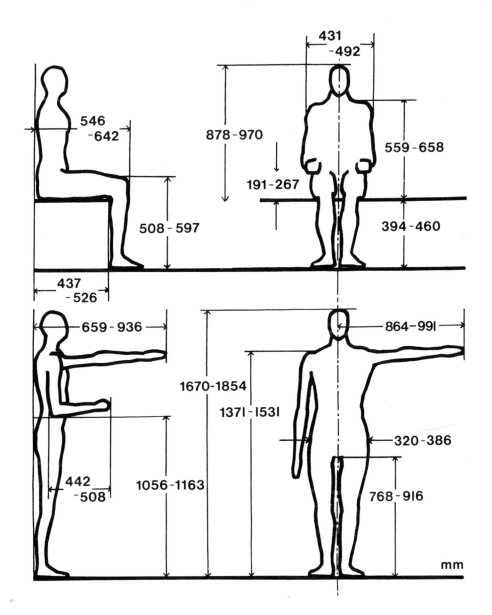

Figure 98
Dimensions of male body given
in 5th and 95th percentile values

Useful technical terms (p) = prefix (s) = suffix

A —, Ab — (p)	Away from
Acceleration	The rate of increase of a measurement per unit of time
Accretion	The process of growth by external addition to cause increase
Acromion	The highest point on the most distal part of the spine of the scapula
Ad — (p)	To, toward
Adaptation	The ability to make changes in the physiological system in reaction to noted environmental conditions without detracting from mental or physical abilities
Allometry	Differential growth of one part of the body to the remainder. It is because of differential growth rates that the body exhibits changes of form and proportion
A —, An — (p)	not, without
Ante — (p)	Before
Anterior	To the front
Anthropomorph	A conventional human figure: to resemble a human form
Anti — (p)	Against
Articulation	The relationships of two parts by means of a moveable joint
Auto — (p)	Self
Axilla	The armpit
Bi — (p)	Two
Bio — (p)	Life
Biotypology	In biology the study of genotypically identical individuals. In design the distinguishing of groups of people with common physiques or physical proportions
Canthus	The angle at the junction of the eyelids
Caudal	Relating to the tail
Circum — (p)	Around
Circumduction	A movement at a joint; the limb revolving round an imaginary axis so as to describe a cone
Concentric	Concentric contraction. Muscle tension overcoming resistance and shortening
Contra — (p)	Against, opposed
Di — (p)	Two
Dia — (p)	Through
Dimorphism	The occurrence and display of two forms distinct in proportion and structure amongst animals of the same species

Discrete	Separate
Distal	Furthest from the heart or from the median line. At the far end
Dorsal	Belonging to the back, directed backwards
Dynamic	Relating to a force and movement
Dys, — (p)	Bad, difficult
Eccentric	Eccentric contraction. Resistance overcoming muscle tension; muscle can lengthen
Ecology	The study of the home conditions
Ecto — (p)	On outside
Ectomorphic	The third component of Sheldon's somatotype system
Endo -- (p)	Inside
Endomorphic	The first component of Sheldon's somatotype system
Endocrine	See Glands
Environment	The envelope of conditions in which a person lives or works; a geographic or industrial location reckoned in terms of temperature, humidity, air pressure or light and all other external stimuli
Epi — (p)	Out, away from
Epiphysis	A separate centre of ossification from the main portion of a growing bone, from which it is isolated by cartilage
Ergonomics	The study of human mental and physical performance in any work situation in terms of efficiency and competence. Additionally the design of ideal working environments and the allocation of tasks in a man and machine situation
Eversion	To turn outwards
Ex — (p)	Out, away from
Factor	A quantity or circumstance that produces a result
Frankfurt plane	An imaginary plane passing through the skull from the lower orbits of the eyes to the upper rim of the auditory aperture. In body measurement maintained in the horizontal plane
Fossa	Meaning a ditch. In anatomy applied to a variety of depressions
Fulcrum	The pivot point of a lever which is favourable for applying force
Gene	The units of inheritance which help to form the characters of an individual
Genetics	The study of heredity
Genotype	The constitutional inheritance which reacts with the environment to mould the phenotype

Gerontology	The branch of medical science that deals with the treatment of the aged
Glands	Organs which contain secretory cells. The endocrine (or ductless glands) are of no standard type, but are spread throughout the body and play a vital part in the functions that promote growth and development
Glabella	Most forward point (in the median plane) of the forehead, between the eyebrows and just above their level
Habitat	The locality in which an organism can live and leave descendants
Hemi — (p)	Half
Hetero — (p)	Varied, different
Histo — (p)	Tissue
Homeo — (p)	Similar, alike
Homeostasis	The ability to continue a steady or optimum condition of life by processes of internal self-regulation or adjustment. Self-regulation
Homo	A genus of primates, of the order mammals
Homo-sapiens	The human animal, the only living representative of the genus Homo
Hyper — (p)	Above, excess
Hypo — (p)	Below, deficient
In — (p)	Into, not
Infra — (p)	Below, under
Inter — (p)	Between
Intra — (p)	Within or inside
Inversion	To turn inwards
Ipse — (p)	Same
Iso — (p)	Equal
Isometric	Static contraction. Muscle develops tension with no body parts moved
Isotonic	In physiology the term for both concentric and eccentric contraction
Juxta — (p)	Next to
Kinesiology	The mechanics of the motion of the human body
Kyphosis	Abnormal curvature of the spinal column with lumbar portion dorsally displaced. Hunch-backed
Linea aspera	A prominent ridge on the back of the femur bone for the attachment of muscles
Locomotion	The power of an animal to exert force progressively on the linbs and produce motion
Lordosis	Abnormal curvature of the spinal column with lumbar portion ventrally displaced. Sway-backed

Macro — (p)	Large
Mammal	A class of vertebrates. Distinguished by the female suckling young, a unified lower jaw and the possession of body hair at some period
Median	In the centre, nearer or towards the centre or middle. The mid-sagittal plane
Mega — (p)	Great
Menton	The lower edges of the chin in the median plane. Mentum, the chin
Meso — (p)	Middle
Mesomorphic	The second component of Sheldon's somatotype system
Metabolism	The bio-chemical and bio-physical processes by which life is maintained in the body, and by which energy is released
Meter — (s)	Measure
Mono — (p)	One, single
Morpho	Shape, form
Morphogenotype	The individual form of physical constitution after growth and development. The individual somatotype
Naso — (p)	The nose
Nasion	A point where the nasal bones meet the frontal bone
Neokinetic	Intelligent behavioural learning in man. Learning and behaviour which become habits or conditioned reflexes.
—Oid (s)	Like
—Ole (s)	Small
Optimum	The amount of stimulus that allows all the processes of a function to behave
—Opia (s)	Vision
Opistho — (p)	In composition, behind
Ossification	The production of bone; either from cartilage or embryonic tissue
—Ossis (s)	A condition, a process
Paidology (Paedology)	The study of growing children
Parameter	The construction of a convenient variable in which other variables can be expressed
Parallax	The difference in direction, or shift in the apparent position of a body due to a change in position of the observer
Pathology	The branch of medical science that deals with the causes of disease and the changes they produce in the body
Phenotype	The form that is appearing in childhood and which is due to both genetic and environmental influences

Photogrammetry	Photographic methods used in anthropometry using single or multiple camera techniques
Plantigrade	The act of using the whole of the sole of the foot in walking
Poly — (p)	Many
Pre — (p)	Before, in front
Prehensile	The capability of grasping or gripping
Pro — (p)	Before, giving use to
Proprio — (p)	One's own
Prone	Having the front portion of the body downwards
Proportion	The relationship between two quantities or sizes. The relationship between the part and the whole
Post — (p)	Behind, after
Proximal	Nearest to the heart or to the median line. At the near, inner, or attached end
Puberty	The change of form and function that takes place as a child enters maturity
Retro — (p)	Backward, behind
Sclero — (p)	Hard
Semi — (p)	Half
Scye	The sleeve hole
Sexual dimorphism	The characteristic differences between males and females of the same species
Soma	The body
Somatotype	Body type, figure type
Somatotyping	The classification of body builds invented by Sheldon
Somatology	The science of the human body
Static	Bodies at rest, or the equilibrium of forces when no motion is produced
Stereo — (p)	Solid, three dimensional
Steriophoto-grammetry	Sterioscopic vision camera technique applied to determining depths of body contours as well as linear dimensions
Stimulus	That which can effect or excite the activity or function of an organ to produce a material reaction
Sub — (p)	Below
Super — (p)	Above
Supine	Having the front portion of the body upwards
Syndesmology	The description and study of articulations in anatomy
Synergy	Muscles working together
Tactile	The sense of touch. Perception and belief by touch
Tragion	The point located by the notch in the ear

| | cartilage above and forward of the ear hole |
| Tragus | Small prominence at the entrance of the external ear |

Bibliography

Anthropometry and Photogrammetry

Bolton, C B, Kenward, M, Simpson, R E, and Turner, G M, 1973, *An anthropometric survey of 2000 Royal Air Force Air Crew, 1970/71*. 1 A M Report No. 531, Royal Aircraft Establishment and the Royal Air Force Institute of Aviation Medicine, Farnborough

Chaffe, J W, 1961, *Andrometry: A practical application of coordinate anthropometry in Human Engineering*. Report FZY — 012, Convair Division of General Dynamics Corporation, Fort Worth, Texas, USA

Churchill, E, Kikta, P, and Churchill, T 1977, *Intercorrelations of Anthropometric Measurements: A source book for USA data*, AMRL — TR— 77 --2, Aerospace Medical Research Laboratory, Wright Patterson Air Force Base, Ohio, USA

Croney, J E, 1977, *An Anthropometric Study of Young Women Fashion Students Including a Factor Analysis of Body Measurements*, Man (NS) 12, 484–496

Herron, R E, 1972, *Stereophotogrammetry in Biology and Medicine*, Biosteriometrics Laboratory, Texas Institute for Rehabilitation and Research, Houston, Texas, USA

Hertzberg, H T E, 1954, *Anthropometry of Flying Personnel, USAF*, Wright Patterson Air Force Development Centre, TR. 53 — 32, Wright Patterson Air Force Base, Ohio

Hertzberg, H T E, Dupertius, C W and Emmanuel, I, 1958 *Stereophotogrammetry as an Anthropometric Tool*, WADC, TR 58 — 67(AD150 — 964) Wright-Patterson Air Force Base, Ohio, USA

Karpino, B D, 1958, 'Height and Weight for Selective Service Registrants Processed for Military Service During World War II', *Human Biology* No 30

Roberts D F, 1960, 'Functional Anthropometry of Elderly Women', *Ergonomics* No. 3

Tanner, J M and Weiner, J S 1949, 'The Reliability of the Photogrammetric Method of Anthropometry, with a Description of a Miniature Camera Technique', *American Journal of Physical Anthropology*. 7:2 145 - 186 (NS)

Thompson, O, Barden, J O, Kirk, N S, Mitchelson, D L, Ward, J S, 1973, *Anthropometry of British Women*, The Institute for Consumer Ergonomics Ltd, University of Technology, Loughborough, Leicestershire

Applied anthropometry

AGARD, 1955 *Anthropometry and Human Engineering*, Butterworth Scientific Publications

Akerblom, B, 1948, *Standing and Sitting Posture*, A B Nordiske Bokhandelm, Stockholm

Alexander, M and Clauser, E E, 1965, *Anthropometry of Common Working Populations*. AM RL -- TR - 65 —73, Wright-Patterson Air Base, Ohio: Aerospace Medical Research Laboratories, USAF

Barkla, D M, 1964, '*Chair Angles, Duration of Sitting and Comfort Ratings*', *Ergonomics* No. 7

Barter, J T, 1957, *A Statistical Evaluation of Joint Range Data*, Aero Medical Laboratory, USAF Wright Aid Development Center, TR — S7 — 311

Branton, P, 1966, *The Comfort of Easy Chairs*, FIBRA Report No. 22

Branton, P 1966, *Seating in Industry*, Ministry of Technology

British Standards Institution 1958, '*Anatomical, Physiological and Anthropometric Principles in the Design of Office Chairs and Tables*', BS 3044

British Standards Institution 1959, '*Anthropometric Recommendations for Dimensions for Non-Adjustable Office Chairs and Tables*', BS 3079

British Standards Institution 1959, '*Anthropometric Recommendations for Dimensions of Office Machine Operators*' BS 3404

Carlsoo, S, 1961, '*Muscle load in working postures*', *Ergonomics* No. 4

Chapanis, A R E, 1959, *Research Techniques in Human Engineering*, John Hopkins Press

Chapanis, A R E, 1947, *Lectures on Man and Machines. An Introduction to Human Engineering*, John Hopkins University Report 166 — 1 — 19

Clapham, J R C, and Durdin, J, 1954, '*Chairs and Sitting*', Proceedings Ergonomics Research Society No. 2

Clements, E M, Baverstock, M and Pickett, K, 1955, *The Anthropometric Considerations which Underlie Sizes of Children's Clothing*, University of Birmingham, Department of Anatomy, Report No. 1

Damon, A, and McFarland, R A, 1955, '*The Physique of Bus and Truck Drivers: With a Review of Occupational Anthropology*', *American Journal of Physical Anthropology*, No. 13

Darcus, H D and Weddell, A G M, 1947, '*Some Anatomical and Physiological Principles Concerned in Seats for Naval War Weapons*', *British Medical Bulletin* No. 5

Dempster, W T, 1955, *The Anthropometry of Body Action*, American New York Academy of Science No. 63

Durand, V and Grandjean, E, 1963, '*Sitting Habits of Office Employees*', *Ergonomics* No. 6

Fisher, M B and Birren, J E, 1946, '*Standardization of a Test of Hand Strength*' *Journal of Applied Psychology* No. 30

Floyd, W F and Roberts, D F, 1958, '*Anatomical and Physiological Principles in Chair and Table Design*' *Ergonomics* No. 2

Floyd, W F and Welford, A T, 1954, *Symposium on Human*

Factors in Equipment Design, H K Lewis and Co Ltd

Hertzberg, H T E, 1955, *Some Contributions of Applied Physical Anthropology to Human Engineering* American New York Academy of Science No. 63

Hertzberg, H T E, Emmanuel, L, and Alexander, M, 1956, *The Anthropometry of Working Positions: A Preliminary Study*, WADC -- TR — 54 — 520(AD110 — 573) Wright-Patterson Air Force Base, Wright Air Development Center, USA

Hick, W E, 1952, *'Why the Human Operator?'* Transactions of the Society of Instrument Technology

Hooton, E A, 1955, *A Survey in Seating*, Harvard University

Hugh-Jones, P, 1947, *The Effect of Limb Position in Seated Subjects: Their Ability Utilize the Maximum Contractile Force of the Limb Muscles'*, *Journal of Physiology* No. 105

Hunsicker, D A and Creey, G, 1957, *Studies in Human Strength*, Research Quarterly of the American Association for Health, Physical Education and Recreation No. 28

Jones, J C and Thornley, D G, (editors), 1963, *Conference on Design Methods*, Pergamon Press

Keegan, J J, 1954, *'Anthropometry of Chairs'*, Architectural Design

Keegan J J, 1953, *'Alterations in the Lumbar Curve Related to Posture and Seating'*, *Journal of Bone and Joint Surgery* No. 35A

Keegan, J J, 1962, *'Evaluation and improvement of Seats'*, *Industrial Medicine and Surgery* No. 31

Kemsley, W F F, 1957, *Women's Measurements and Sizes*, London: Joint Clothing Council Ltd, Her Majesty's Stationery Office

Krogman, W M and Johnson, F E, 1963, *Human Mechanics*, Aerospace Medical Research Laboratories, USAF Wright Patterson Air Force Development Center, AMRL — TDR — 63 — 123

Martin, W E, 1954, *The Functional Body Measurements of School Age Children*, Chicago National Service Institute Publication

McConville, J T and Hertzberg, H T E, 1966, *A Study of One Handed Lifting*, Aerospace Medical Research Laboratories, USAF Wright Patterson Air Force Development Center, AMRL — TR — 66 — 17

McCormick, E J, 1957, *Human Engineering*, McGraw-Hill Book Co Inc

McFarland, R A, 1958, *'Anthropometry in the Design of Drivers Workspace'*, *American Journal of Physical Anthropology* No. 16

Morant, G M, 1954, *'Body Size and Work Spaces'*, Proceedings Ergonomics Research Society No. 2

Morgan, C T, 1963, *Human Engineering to Equipment Design*, McGraw-Hill Book Co Inc 2nd ed, Vancott, H P and Kinkade, R G, (eds) 1972, US Government Printing Office, Superintendent of Documents, Washington, DC

Murrell, K F H, 1965, *Ergonomics*, Chapman and Hall

O'Brian, R and Shelton, W C, 1941, *Women's Measurements for Garments and Pattern Construction*, Department of Agriculture, Washington, DC, Publication No. 454

O'Donovan, B, 1961, *'Seating Dimensions Theory and Practice'*, *Design*

Parnell, R S, 1951, *Physical Body Measurements*, Transactions Assistant Industrial Medical Officers

Roberts, D F, 1962, *Human and Inhuman Factors in Office Design*, Paper for Royal Society of Health

Roberts, D F, 1956, *'Industrial Applications of Body Measurements'*, *American Anthropology* No. 58

Roebuck, J A, Kroemer, K H E, and Thomson, W G, 1975 *Engineering Anthropometry Methods*, John Wiley & Sons, London

Renbourn, E T, 1971, *Physiology and Hygiene of Materials and Clothing*, Merrow, London

Taylor, C I and Boelter, L M K, 1946, *'Biotechnology, A New Fundamental in the Training of Engineers'*, *Science* No. 105

Welford, A T, 1960, *'Ergonomics of Automation'*, Department of Scientific and Industrial Research HMSO

Whitney, R J, 1958, *'The Strength of the Lifting Action in Man'*, *Ergonomics* No. 1

Woodson, W E and Conover, D W, 1970, 2nd ed, *Human Engineering Guide to Equipment Designers* University of California Press, Berkeley, California

Growth, ageing and physique

Barton, W H and Hunt, E E, 1967, *'Somatotype and Adolescence in Boys: A Longtitudinal Study'*, *Human Biology* No. 4

Belbin, R M, 1955, *'Older People and Heavy Work'*, *British Journal of Industrial Medicine* No. 12

Birren, J, 1959, *Handbook of Ageing and the Individual*, University of Chicago Press

Bourne, G H, 1961, Editor, *Structural Aspects of Ageing*, Pitman

Brozek, J, Hunt, E E and Skerji B, 1953, *'Subcutaneous Fat and Age, Changes in Body Build and Body Form in Women'* *American Journal of Physical Anthropology*, No. 11

Edwards, D A, 1951, *'Differences in the Distribution of Subcutaneous Fat with Sex and Maturity'*, *Clinical Science* No. 10

Edwards, D A, 1959, *'Observations on the Distribution of Subcutaneous Fat'*, *Clinical Science* No. 9

Gallagher, J R and Seltzer, C C, 1946, *'Somatotypes of an Adolescent Group'*, *American Journal of Physical Anthropology*, No. 2

Jones, H E, 1947, *'The Relationship of Strength to Physique'*, *American Journal of Physical Anthropology* No. 5

Kemsley, W F F, 1952, *'Bodyweight at Different Ages and*

Weights', Annals Eugenics

Sheldon, W H, Dupertius, C S and McDermott, E, 1954, *Atlas of Men*, Harpers, New York

Sheldon, W H and Baker, W B, 1940, *The Varieties of Human Physique*, Harpers, New York

Tanner, J M, 1964, *The Physique of the Olympic Athlete*, G Allen and Unwin Ltd

Tanner, J M, 1955, *Growth at Adolescence*, second edition, Blackwell Scientific Publications

Welford, A T, 1958, *Ageing and Human Skill*, Oxford University Press for Nuffield Foundation

Human biology

Harrison, G A, 1977, Ed, and others. Second edition *Human Biology*, Oxford University Press

Weiner, J S, and Lourie, J A, 1969, *Human Biology, A Guide to Field Methods*. Handbook No. 9, Blackwell Scientific Publications

Statistics

Bailey, N T J, 1975, *Statistical Methods in Biology*, Hodder and Stoughton, London

Ehrenberg, A S C, 1975, *Data Reduction*, John Wiley & Sons, London

Hill, A B, 1971, *Principles of Medical Statistics*, Oxford University Press, London

Moroney, M J, 1976, *Facts from Figures*, Penguin Books Ltd, Harmondsworth, Middlesex, England

Reichmann, W J, 1961, *Use and Abuse of Statistics*, Penguin Books Ltd, Harmondsworth, Middlesex, England

Vann, E, 1972, *Fundamentals of Biostatistics*, Heath and Company, Lexington, Massachusetts and London

Index